A SHORTCUT
THROUGH TIME

George Johnson is a science writer for the *New York Times*. He is a former Alicia Patterson Fellow, a finalist for the prestigious Aventis Prize, and a recipient of the Science Journalism Award from the American Association for the Advancement of Science. His books include *In the Palaces of Memory, Fire in the Mind* and *Strange Beauty*.

George Johnson

A SHORTCUT THROUGH TIME

The Path to the
Quantum Computer

VINTAGE

Published by Vintage 2004

2 4 6 8 10 9 7 5 3 1

Copyright © George Johnson, 2003

First published in the USA in 2003 by
Alfred A. Knopf

First published in Great Britain in 2003 by
Jonathan Cape

Vintage
Random House, 20 Vauxhall Bridge Road,
London SW1V 2SA

Random House Australia (Pty) Limited
20 Alfred Street, Milsons Point, Sydney
New South Wales 2061, Australia

Random House New Zealand Limited
18 Poland Road, Glenfield,
Auckland 10, New Zealand

Random House (Pty) Limited
Endulini, 5A Jubilee Road, Parktown 2193,
South Africa

The Random House Group Limited Reg. No. 954009
www.randomhouse.co.uk/vintage

A CIP catalogue record for this book
is available from the British Library

ISBN 0 09 945217 0

Papers used by Random House are natural, recyclable
products made from wood grown in sustainable forests.
The manufacturing processes conform to the environ-
mental regulations of the country of origin

Printed and bound in Great Britain by
Bookmarque Ltd, Croydon, Surrey

In memory of my father, Dr. Joseph E. Johnson,
November 16, 1917 – March 29, 2001

Contents

Preface: Inside the Black Box

I remember with some precision when I began believing that there is nothing so complex that a reasonably intelligent person cannot comprehend it. It was a summer day, when I was fifteen or sixteen, and my best friend, Ron Light, and I decided that we wanted to understand how a guitar amplifier works. We both played in a mediocre 1960s-era garage band. While Ron went on to become a fairly accomplished guitarist, I was slowly learning that any talent I had didn't lie within the realm of music. Already the aspiring little scientist, I was able to learn enough of the logic of basic harmony theory to execute the mindlessly simple algorithms called bass riffs, and if pressed I could even fire off a bass solo, the dread of concertgoers everywhere. But my approach to the performance was purely intellectual. I didn't have rhythm, or maybe soul.

Poring over the symbols on the circuit diagram of Ron's Fender Deluxe Reverb amplifier seemed infinitely more interesting than trying to read music. I wanted to know what that impressively convoluted blueprint really meant, how electricity flowing through the labyrinth of wires and components could cause the tiny vibration of a guitar string to be multiplied so many times that it rocked the walls of the living room, incensing the neighbors into calling the police.

This was still the era of the vacuum tube, before those wonderful glowing glass envelopes were replaced by coldly efficient transistors and microchips. Electronics was pretty simple to understand. I had already learned some basics from *The Boys' Second Book of Radio and Electronics* and the guide for the Boy

Scout electricity merit badge (the colorful embroidered patch was decorated with a human fist clutching zigzag lightning bolts). In a typical circuit, there were resistors that, true to their calling, resisted electricity, pinching the flow of electrons. There were capacitors, also aptly named, that stored electrical charges. There were tightly wound coils of copper wire called inductors that would invisibly hold energy in the form of electromagnetic fields. Finally there were the vacuum tubes themselves, mysterious pockets of illuminated nothingness inside of which the actual amplification took place.

At first the detail and complexity of the schematic, showing how all these parts fit together inside the Fender's vinyl-covered wooden cabinet, were overwhelming. I could feel my mind start to shut. But with the help of some slightly more advanced books from the Albuquerque Public Library, I realized that I was taking the wrong approach. The trick was to break down the diagram into pieces, master each one, and then put them back together again.

Before long I could place my finger on the diagram and follow the path of the vibrating electrical signal—a replica of the sound of the twanging guitar or the thumping bass—as it traveled through the maze of squiggly lines. Each of the mysterious vacuum tubes, I came to see, was nothing more than a lever. The minuscule fluctuating voltage emerging from the guitar was fed to the first tube, where it was used to operate a gate that controlled a second, much bigger voltage. What resulted was a larger copy of the original signal. This was sent on to the next tube and leveraged again. Step by step the undulating swings were transformed into ones wide enough to move the cone of the loudspeaker . . . which would ripple the air and shake your eardrums and stimulate the auditory nerve—a kind of neural guitar pickup that turned the vibrations back into electricity again, input for the brain.

Here was the best part: it was only incidental that this seesawing cascade was being pushed and pulled with electricity. One could imagine a completely hydraulic system where the signal was carried by tubes of vibrating fluid moving a series of larger and larger mechanical diaphragms. In theory you could make the fulcrums from gears and pulleys or wooden spools and string.

There were good reasons for not using these clunkier technologies. The delicate, nearly weightless electrons could be controlled with a finesse not possible with mechanical parts. The point of the mental exercise was not to make hydraulic guitar amps, but to abstract the concept of audio amplification beyond its incidental underpinnings. Peeled away from one particular embodiment, the Fender amplifier, the idea revealed itself as simple and profound. I didn't need or want to understand amplification with the razor-sharp acumen of an engineer. I didn't care about being able to perform a mathematical analysis of the circuits or to understand the finer nuances of esoteric concepts like the "hysteresis" of a transformer or the "mutual conductance" of a tube. I just wanted a gut-level feel for what those electrical parts were doing.

By the time I was in college, I could zero in on a malfunctioning circuit and repair it. I could add tubes to the output stage of a lowly Deluxe Reverb, turning it into a more powerful and expensive Super Reverb. I was amazed that I could get so far with just the broad outlines of understanding.

Then I took on television.

This turned out to be a little harder, but I soon found that you could adjust the focus of your curiosity up and down, from fine to fuzzy. You may have neither the time nor the inclination to grasp a video circuit in great detail. Suppose you have gleaned from your reading that some conglomeration of components called an oscillator—a kind of electrical spring capable of rhythmically plucking itself—produces a vibrating electrical signal, which is fed into an electromagnet, a yoke of wire coiled around the neck of the picture tube. The result is a fluctuating field that drives a beam of electrons sweeping back and forth, up and down, painting an image on the phosphor screen.

Now just draw a line around the appropriate squiggles on the schematic diagram and treat everything inside as a black box. Color it solid black, if you'd like, for from now on you will ignore whatever is within. You can take it on faith that, given a certain input, the circuitry produces a certain output. Later on, if you like, you can pry off the lid and zoom in closer for a more detailed view. Or you can pan outward, lumping the circuitry into bigger and cruder chunks. Most people look at the whole TV as one big black box that takes signals from the air and magically turns them into sound and pictures. Any device, no matter how complex, can be understood on many different levels of abstraction.

I didn't appreciate back then that I was already approaching the world like a science writer (with an audience of one). Whether you are taking on molecular biology, cosmology, or dendrochronology (we'll leave that word a black box), you are learning as you go along. Like a pilot of a plane looking down from on high, you let the minor geographical details blur together, leaving you to concentrate on the most arresting features of the terrain. When you spot some particularly alluring region, a chain of mountains or a convergence of streams, you can swoop down closer for a finer look, but not so close that you become lost in the details, forgetting the lay of the land.

You are not an expert taking stuff you already know and simplifying it for a general audience. You are part of the audience—at first, anyway. As your exploration proceeds, you become an unusually active participant, downloading papers from the Web, scrutinizing the hieroglyphics, trying to glean enough from the introduction and the conclusion to ask a few good questions. Before long you are barraging the scientists with e-mails and telephone calls, then visiting some of them in their labs. But you always maintain a certain distance, a detachment. That is part of your bargain with the reader. You have no axe to grind. The goal is to show how some new discovery looks to an interested outsider, writing for other interested outsiders, using metaphor instead of mathematics.

Over the years I've tried to use this approach to give both readers and myself crash courses in artificial intelligence, the neurobiology of memory, particle physics, and the new science of complexity. A couple of years ago, Kevin Kelly, one of the founders of *Wired* magazine, urged me to take on what may be the hardest task yet: explaining something called quantum computing. I'd written about some of the developments for the *New York Times,* describing how scientists were trying to compute using invisibly small strings of atoms. People who follow science or science fiction have a vague notion that quantum mechanics somehow defies the restrictions of ordinary reality, allowing tiny objects to take quantum leaps from one point to another, without traversing the space in between, or to somehow exist in multiple places. Exploiting such loopholes, a quantum computer would be able to do a vast number of calculations at the same time, solving problems that would be otherwise impossible.

What the world needs, Kelly told me, is a short book (emphasis on "short") that would explain how one of these machines would work. Is this pie-in-the-sky theorizing, or is everything we know about to change?

So I began the cycle anew, the gathering of papers, the dispatching of questions, the visiting of labs. Along the way, I decided to abandon some of the usual tools of the trade. Much as I like to write and read narratives that weave together the science and the personalities of the scientists, I decided that this book would be a little different. The story would be driven entirely by the ideas. I wouldn't concern myself with the way one mathematician's mustache wiggles when he talks or the ghostly appearance of a certain nocturnal British physicist who believes that each of the myriad calculations of a quantum computer takes place in a different universe. Sticking to the ideas, I hoped, would impart a crispness to the book and speed passage from beginning to end.

Like any kind of writing, science writing involves spinning an illusion. All the hard intellectual work—digging through the piles of papers and reference books, reading the same paragraph a dozen times before firing off yet another e-mailed plea for clarification—all this takes place behind a curtain, carefully hidden from view. What ultimately emerges, Oz-like, is a narrator who speaks with the resounding, omniscient voice of authority, a being seemingly born with encyclopedic knowledge instantly retrieved and dispensed. And that, of course, is a fiction. A friend who was reading one of my books once asked, "Do you actually know all those things you put in there, or do you have to look them up?" She was relieved to hear that what seemed a smooth flow of effortless erudition was haltingly cobbled together after multiple trips to the library. "But isn't that cheating?" she said. I think she was kidding, but sometimes I feel that way. I leaf through the index of a book I've written and marvel, "What in the world did I have to say about Plotinus or Aristotle?"

This time I have tried not to cheat. In the pages that follow, the surface has been left somewhat translucent, offering dim glimpses of the man behind the curtain fumbling at the controls—straining to grasp an idea with an imprecise metaphor, only to discard it for another with a tighter fit, closing in on an airy notion from several directions, triangulating on approximate truth.

I've also tried to resist the temptation to say too much. Fascinating as they are, many of the more tangential details of quantum mechanics and computer science—the two threads that wind through this story—will remain wrapped inside their boxes. We are operating here on a need-to-know basis. (Those who want to look deeper can refer to my annotations at the end of the book, called "The Fine Print," a section that can also be taken as a gloss on the nature and limitations of science writing.)

More than ever, I want the reader to feel that we are both on the same side—outsiders seeking a foothold on the slippery granite face of a new idea. My guiding light has been a statement by the writer Alan Lightman about what makes a good essay. I think it applies as well to a good nonfiction book: "For me, the ideal essay is not an assignment, to be dispatched efficiently and intelligently, but an exploration, a questioning, an introspection. I want to see a piece of the essayist. I want to see a mind at work, imagining, spinning, struggling to understand."

A Shortcut Through Time

Prologue:
The Road to Blue Mountain

Just beyond the narrow passage separating two mesas where Otowi Bridge spans the Rio Grande, New Mexico State Road 502, one of the most spectacular highways in the country, begins its steep ascent up Pajarito Plateau, home of Los Alamos National Laboratory. For a brief period during World War II, the inhabitants of this sky island possessed the most concentrated package of man-made energy in the world: the 20-kiloton nuclear bomb that melted sand into glass at the desert test bed called Trinity Site.

Though still the country's premier weapons lab, Los Alamos no longer builds warheads. It studies them. And the most powerful instrument in the scientists' high-altitude armory is not a bomb but a supercomputer called Blue Mountain.

Housed in an expansive room about a quarter of an acre in size and hidden away inside the main cluster of laboratory buildings, Blue Mountain is one of the mightiest calculators in the world. The computer sitting on your desktop is probably powered by a single high-speed processing chip whose millions of tiny switches click back and forth, chattering 1s and 0s, several hundred million times a second. Each of the 384 towering cabinets that makes up Blue Mountain contains 16 such chips. The result is a machine in which a total of 6,144 processors can join in parallel to collaborate on one horrendously complex problem: simulating a nuclear explosion. In the post–Cold War era, when testing nuclear bombs in the atmosphere or even below the ground has been banned by treaty, the Energy Department is trying to test them instead "in silico"—by mimicking their explosions in Blue Mountain's 1.5 trillion bytes of silicon memory chips.

The computer's keepers proudly reel off its formidable specs. The thousands of processors are laced together with some 500 miles of fiber-optic cable, wire that carries pulses of light. The machinery consumes 1.6 megawatts of electrical power requiring 530 tons of cooling capacity. The result is called a 3-teraops computer, meaning that it performs 3 trillion mathematical operations a second.

With a little practice, a dexterous individual might perform a single mathematical operation, multiplying 2×2 with a pocket calculator, in two or three seconds. Suppose you could punch in those numbers in a single second. There are about 3 billion seconds in a century, so it would take you a thousand centuries to do what Blue Mountain can do in one second. Going back that far would bring you to the late Middle Pleistocene epoch when *Homo sapiens* was emerging.

Though awed by Blue Mountain when it was installed in 1998, Los Alamos's weapons experts soon began finding it agonizingly slow, a feeling with which any PC owner, saddled with last year's state-of-the-art electronics, can empathize. When they recently ran a calculation intended to reproduce a millionth of a second of a nuclear blast, so much data had to be processed that it took four months of number crunching. Replacing real explosions with artificial tests will require a far more powerful machine.

Not far from Blue Mountain, the laboratory recently put the finishing touches on a computer so large that it is surrounded by its own three-story, 303,000-square-foot building. Containing almost twice as many processors as Blue Mountain, each running five times faster, the result is a 30-teraops machine. It is called simply Q. The name shimmers with connotations. Q is the name of a dimension-hopping alien in *Star Trek: The Next Generation,* as well as a James Bond character who kept his clientele well stocked with the most sophisticated exotic spying gadgetry. Q is the last letter in Compaq, the company that made the hardware for the supercomputer, and the name of a high-level government security rating is a Q clearance. Spewing heat through its cooling towers, the 7-megawatt calculating factory will perform as many computations in a second as a lightning-fingered button pusher could in a million years.

The floor holding Q is four times larger than the one for Blue Mountain, a full acre. At first the computing machinery will occupy only about half the space. But as the lab builds toward its goal, many years away, of a 100- or even 150-teraops computer, the designers expect the vacant real estate to be overtaken with cybernetic sprawl. In a honeycomb of laboratories surrounding the building's core, legions of scientists will study Q's ruminations. For maximum dramatic effect they will be able to step inside large wraparound theaters, watching in slow motion as a simulated explosion unfolds around them. In the eye of the nuclear tornado.

Just a short walk from the computing center—until recently a cacophony of beeping backhoes, whining metal saws, and other industrial noise—two young physicists, Manny Knill and Raymond Laflamme, have been taking a quieter approach to supercomputing. In a nondescript brown stucco building on the outskirts of the main laboratory complex, they are programming a computer too tiny to see, even with a microscope: a single molecule strung together from a dozen atoms.

Without committing too much conceptual violence, an atom

can be thought of as a spinning top. Depending on whether it is rotating clockwise or counterclockwise, it can be used to indicate a 0 or a 1, the two binary digits, or "bits," in the universal language of computation. By surrounding molecules with an intense magnetic field and flipping their atoms with high-frequency radio waves, the scientists can manipulate short strings of bits, carrying out simple calculations. They recently set a record, using seven of the molecule's atoms as the abacus beads in what is called a quantum computer. The next milestone is ten.

Considering that Intel's Pentium 4 chip contains 40 million switching elements, ten sounds like a rather paltry number. But when switches are this small—a single atom in size—they obey the rules of quantum mechanics.

Conveying what that means is the goal of this book. But for now, a rough approximation will do: In the tiny spaces inside atoms, the ordinary rules of reality (ordinary, anyway, from the human perspective) no longer hold. Defying all common sense, a single particle can be in two places at the same time. And so, while a switch in a conventional computer can be either on or off, representing 1 or 0, a quantum switch can paradoxically be in both states at the same time, saying 1 *and* 0. If that seems impossible, don't despair. The physicists find it as puzzling as the rest of us do. It is just the way nature seems to work.

Why this quantum ambiguity is a good thing and not just a source of computational confusion becomes a little clearer when you consider what you could do with two quantum switches, each with this same schizophrenic quality. At any one time, two ordinary switches can be in one of four different configurations. Both can be off or both can be on: 00 or 11. Or they can be in opposite states: 01 and 10. In the quantum realm, the either-or rule breaks down. If one quantum switch can simultaneously be in two positions, then two quantum switches can simultaneously be in four: 00, 01, 10, and 11. Three switches can be in eight states: 000, 001, 010, 011, 100, 101, 110, and 111—all at the same time.

Therein lies the source of the power. With three ordinary switches, each set to 1 *or* 0, you could store any one of these patterns, which are the binary encodings of the numbers 0 through 7. Three quantum switches, each crying out both 1 *and* 0, can hold all eight numbers at once. Hard to believe, but true. At the very least this extraordinary phenomenon could lead to ways of packing vastly more information into very small places. But that is just the beginning of the possibilities. Suppose you take this triplet of quantum bits and use it as the input for a computation—dividing by two, or taking a square root. Since the string simultaneously holds eight numbers, you would be doing eight calculations at once.

Don't worry yet about the specifics of performing such a computational juggling act. Think of the string of atoms—the data—as passing through a black box. Going in is a string representing all the numbers to process; coming out is one representing all the results.

And what if you had four atoms? Each time you add an atom to the chain, the number of patterns that can be stored and processed doubles in size: 4 atoms can hold 16 patterns, 5 atoms can hold 32 . . . then 64, 128, 256. By the time you reach 10 atoms, the total is 2 to the 10th power, or 1,024. Want to know the square root of every number from 1 to 1,000? Just load them all onto a row of 10 atoms, perform a single calculation, and you instantly have all 1,000 answers. No one has yet pulled off such a delicate feat, but nothing in the laws of physics seems to prevent it.

With 13 atoms, doubling, doubling, and doubling again, you have a device that can do 2^{13} or 8,192 parallel calculations, surpassing Blue Mountain's mere 6,144. To match the power of the new 30-teraops monster, just add one more atom and double the number to 16,384 calculations, all of which can be carried out in tandem. Such is the power of quantum computation. Perhaps this is the machine that should rightly be called Q.

This comparison between big Q and little Q is not quite exact. The 13 atoms of the molecular computer would be processing data in 13-bit chunks. The new supercomputer bites off data 64 bits at a time. So make the row of quantum switches longer: 64 atoms. The resulting computer would perform 2^{64} or 18,446,744,073,709,551,616 simultaneous calculations. That's 18 quintillion—a trillion repeated one million times.

For a supercomputer like the new one at Los Alamos to do that, it would need millions of trillions of processors. And so, all things being equal, it would occupy 750 trillion acres—roughly a trillion square miles. It wouldn't fit on the planet. The surface of the Earth is just 200 million square miles, so a supercomputer as powerful as the invisible 64-atom quantum calculator would fill the surfaces of 5,000 Earths, assuming you could figure out a way to operate equipment on ocean-floating platforms.

Or you could use a single molecule.

In computer science, the word "impossible" was long ago replaced by a slightly milder one: "intractable." Programmed to play tic-tac-toe, a computer will always win (or, depending on the wiliness of its opponent, tie). At each step of the game, the machine can rapidly sift through all the possible ways the playing might unfold and pick the move that will leave it with the strongest advantage. It will make no mistakes. Doing the same for chess is not impossible, though it might as well be. Completely searching the maze of potential chess plays would take eons, even for a bank of Blue Mountains. The problem is intractable for any conceivable digital computer. (Machines still win at chess because they can rapidly consider far more scenarios than a human player can. It is not necessary to exhaustively analyze every possible variation of the game.)

Breaking a very long number into its factors—the smaller numbers that can be multiplied together to produce the larger one—is also widely believed to be an intractable problem. One possibility after another must be tried ($2 \times 2, 2 \times 3, 2 \times 4 \ldots$) until the answer emerges. The task is easy for small numbers. For numbers hundreds of digits long, there is not world enough or time. Barring an unexpected mathematical breakthrough, factoring numbers hundreds of digits long would take the fastest supercomputers billions of years. Faced with that kind of burden, the impressive speedups promised by each new generation of supercomputer become almost meaningless. Governments and corporations depend on this fact to protect some of their most valuable secrets. Their codes are based on the intractability of factoring large numbers.

It has come as something of a shock for cryptographers to realize that with quantum computation all this would change. In the last few years, mathematicians have proved that for factoring, a quantum computer would be fundamentally more powerful than any "conventional" supercomputer—one that does not exploit the strange possibilities of subatomic behavior. Though the speedup is not quite so dramatic, searching through vast repositories of information could also be done in record time. Because a collection of quantum objects—atoms, electrons, photons—can be in multiple states at the same time, they can perform more simultaneous calculations than could ever be done with ordinary mechanical or electronic parts.

Build a quantum computer and problems long dismissed as hopeless would melt away. Imagine tapping a fundamental force of nature, not for the purpose of moving around matter but for moving around numbers—explosions of information. Quantum computing would be to ordinary computing what nuclear energy is to fire.

Half a century has passed since the physicists of the Manhattan Project came to Los Alamos for the legendary effort that led to the nuclear bomb—a testimony to what (for better or worse) can be accomplished when enough intelligence and money are focused in the same locale. Now the United States government is pouring tens of millions of dollars a year into projects like the one at Los Alamos, a trickle that may quickly grow into a stream. Vulnerable codes are as disturbing to nations as vulnerable borders. The possibility of a new kind of computing is both an opportunity and a threat.

At the National Institute of Standards and Technology's laboratory in Boulder, Colorado, scientists are trapping charged atoms in magnetic fields and using them as tiny computer switches—ones that can be on, off, or on and off at the same time. Other researchers are attempting to use photons, particles of light, as the quantum tokens. The California Institute of Technology, the Massachusetts Institute of Technology, the University of California at Berkeley, Stanford University, the University of Michigan, the University of Southern California—all these and many others are studying quantum computing. Outside the United States, the University of Oxford has become a leading research center. In the corporate world, IBM, Lucent, and Microsoft have also been researching the feasibility of using subatomic particles to compute. As with nuclear science in the 1940s, a quiet branch of theoretical physics has leaped to center stage.

The question everyone is asking is how soon a useful machine might be built. In their laboratories, scientists are already demonstrating that the basic ideas of quantum computation are sound. To make machines that can solve problems now considered intractable, quantum computing must be carried out on much larger scales. Many atoms will have to be manipulated, not just the handful that have been tried so far.

The optimists hope that quantum computing today might be approximately where nuclear science was in the late 1930s when fission was discovered in tabletop experiments. Several years later came the Trinity explosion and, before long, a nuclear power industry.

Harnessing quantum computing is expected to take at least that long. But with each small milestone, the possibilities grow. For those of us watching from the sidelines, the development of this new kind of computing may be one of the most engaging and important scientific dramas of the century. The ticket for admission is a rough feel for the basic ideas of computation and quantum mechanics, and a willingness to suspend disbelief.

1 | "Simple Electric Brain Machines and How to Make Them"

I don't know where I first saw the advertisement for the Geniac Electric Brain construction kit, but I knew I had to have one for Christmas. It was the early 1960s, and like a lot of science-crazed kids I was obsessed with the wonderfully outrageous idea of "thinking machines." I devoured the picture stories in *Life* magazine and the *Saturday Evening Post* about the electronic behemoths manufactured by companies like International Business Machines, Univac, and Remington Rand. The spinning tape drives and banks of blinking lights were as exciting to me as the idea of space travel. Two of my favorite books were *Tom Swift and His Giant Robot* and *Danny Dunn and the Homework Machine*—testaments to the eerie fantasy of automating human thought. One day, flipping through one of my favorite magazines—probably *Boys' Life* or *Popular Science*—I stumbled upon an unbelievably tantalizing ad.

"Can you *think* faster than this Machine?"

Below the provocative headline was a picture of the Geniac with its sloping panel bedecked with six large dials and a row of ten bulbs. Who knew what kind of mysterious circuitry was hidden inside?

"GENIAC, the first electrical brain construction kit, is equipped to play tic-tac-toe, cipher and decipher codes, convert from binary to decimal, reason in syllogisms, as well as add, subtract, multiply and divide. . . . You create from over 400 specially designed and manufactured components a machine that solves problems faster than you can express them." Such was the promise of the Oliver Garfield Co., 126 Lexington Avenue, New York 16, N.Y. (This was before zip codes replaced the old numbered postal zones.) To a boy growing up in Albuquerque, the location of this modern Frankenstein laboratory seemed promisingly exotic and far away.

Figure 1.1. The ultimate Christmas gift: a brain in a box

"Send for your GENIAC kit now. Only $19.95. . . . We guarantee that if you do not want to keep GENIAC after two weeks you can return it for full refund plus shipping costs."

There was nothing to lose. I began my lobbying effort, making it clear to my parents that receiving a Geniac was all that mattered to me. Then I waited, my brain charged with the kind of high-voltage anticipation that can only accumulate in someone still in the first decade of life.

Christmas morning I sat on the floor anxiously opening presents, keeping my eye out for one large enough to hold the pieces of an electronic computer. Finally a likely box emerged from behind the tree. I tore off the wrapping.

Decades later I still remember the disappointment I felt as I explored the contents of the cardboard package. The title of the instruction manual was intriguing enough: "Simple Electric Brain Machines and How to Make Them." But how was anyone to carry out such an ambitious project with the meager, humdrum parts that had been supplied?

Digging through the pile, I was crestfallen to find that the bulk of the kit consisted of some decidedly low-tech pieces of particle board called Masonite: a big square one and six smaller round disks, each drilled with concentric patterns of little holes. This was complemented by an assortment of hardware you might find in a kitchen junk drawer or a toolbox in the garage: ten flashlight bulbs and sockets, a battery and battery clamp, a spool of insulated wire, several dozen nuts, bolts, and washers, a bunch of small brass-plated staples (referred to in the typewritten instructions as "jumpers"), and the tools for assembling this detritus into what would supposedly function as a digital computer—a hexagonal wrench for gripping bolts (a "spintite") and a screwdriver.

Finally there was a simple on-off switch, described rather melodramatically in the manual: "This is the switch that enables you to put suspense and drama into your machine; for you set everything the way it should be, then talk about it and explain it, and finally when you have your listener all keyed up and ready, you (or he) throw the switch . . ."

I'd been had. There were no vacuum tubes, no transistors, or capacitors, or resistors—the colorful components I'd found from eviscerating dead radios and TV sets. All I'd gotten for Christmas was a handy-dandy kit for stringing together mindlessly simple circuits of switches and bulbs. The nuts and bolts were to be placed in various holes on the square wooden panel and connected one to the other by wires running underneath. The little metal jumpers were to be inserted into holes in the Masonite disks, the ends bent over to keep them in place. When the disks were attached to the panel, with more bolts and washers, they could be turned to and fro so that the jumpers touched the heads of the bolts, forming connections that caused the lightbulbs to flash on and off. It was all just switches—simple enough for a child to understand.

Reluctantly I opened the manual, published, disconcertingly, in 1955, and saw that it contained the familiar explanations of the wonders of electricity. ("You can think of a battery as a pump, which is able to push electrons, or little marbles of electricity, away from the plus end of the battery and towards the minus end of the battery . . . A flow of electrons is an electric *current*.")

The instructions went on to show how to assemble circuits and switches into various question-answering machines.

1. Whom do you prefer: (a) Marilyn Monroe? or
   (b) Liberace?
2. How would you put a thread in a small hole: (a) wet it?
   or (b) tap it?
3. Would you rather spend a day: (a) shopping on Fifth
   Avenue? or (b) hunting in the woods?

Depending on how you answered these and three other questions (rotating the six circular switches so that they pointed to A or B), the current in the wiring would flow to one of two bulbs, M or F. The result was called a "Masculine-Feminine Testing Machine."

Never mind the musty Eisenhower-era philosophy. Scientifically, the whole thing seemed obvious and dumb. Just changing the paper labels would turn the machine into a tester for, say, whether you were a Jock or a Brain (the 1960s version of "nerd"): "Would you rather spend a rainy afternoon: (a) building a crystal radio? or (b) working out in the gym?" The meaning was all in the eye of the beholder.

As I paged through the manual, other projects appeared

GENIACS®:

SIMPLE ELECTRIC BRAIN MACHINES, AND HOW TO MAKE THEM

Also:
Manual for Geniac® Electric Brain Construction Kit No. 1

Figure 1.2. Design for a thinking machine

slightly more interesting. The switches could be wired to make machines that added numbers. Turn dial A to indicate the first number, turn dial B to indicate the second, and if the copper paths behind the panel had been correctly platted, the lightbulb that came on would be the very one labeled with the proper answer. (And if you made a mistake and didn't feel like redoing the wires, you could just move around the tags.)

Rig the machine another way and you could subtract or multiply. Wire up the "Reasoning Machine" described on page 25 of the guide and turning switch A to indicate "All fighter pilots are bomber pilots" and switch B to "No bomber pilots are jet pilots" would light the bulb for "No fighter pilots are jet pilots." QED. A syllogism (and a reminder that something can be logical but not true).

It was all terribly anticlimactic. Only years later would I realize that an utterly profound idea was slowly insinuating itself into my head: a computer is indeed just a box with a bunch of switches. The Geniac was "semiautomatic," as the manual put it: You had to turn the dials by hand to light the lights. And to "reprogram" the machine, you had to unscrew nuts with the "spintite" and shift the wires around. But suppose that some of the bulbs were replaced with little motors. When activated by the proper combination of settings, the motors could turn switches on other Geniacs, which would cause the dials on still other Geniacs to spin.

All kinds of elaborations were possible. Data could be fed into a glorified Geniac not with clunky wooden rotary dials but with cards or paper tape punched with holes arranged in the proper patterns. Read like Braille by metallic fingers, the holes would cause electrical connections to be made. Instead of flashlight bulbs, the circuits could ignite phosphorescent dots on a video screen—which is really just an array of thousands of tiny lights.

In the computer's modern incarnation, mechanical parts have been replaced by millions of microscopic transistors, minuscule switches etched onto the surfaces of silicon chips. Data are stored as invisible magnetic spots on spinning disks. But the basic idea is the same.

Only remnants of my Geniac survive. Going through a closet recently, I found a box of old electronic junk. There, sitting atop a partially dissected radio chassis and other relics of the vacuum-tube age, were three of the old dials. Two still had labels—"Candidate A: Popularity," "Candidate A: Campaign Effort"—handwritten in black ink on strips of surgical adhesive tape. I had apparently tried to make a machine to predict election outcomes.

I couldn't find the old manual anywhere. The reason I can quote from it (and the advertisement), reigniting old memories, is because I found the information in minutes on the World Wide Web. (I also found a Geniac kit on eBay, which was auctioned off for more than $400, twenty times its original price.) I'm still astonished by the speed with which data course through the great skein of computers called the Internet. But it is comforting to know that for all its complexity, the Net is just a bunch of Geniacs chattering at each other, twisting each other's dials.

On page 37 of "Simple Electric Brain Machines and How to Make Them" came what was intended to be the climax: instructions for how to make a Geniac that played tic-tac-toe. But looking back years later, it's clear to me that the real meat of the book was an innocent-sounding statement way back on page 3: "The kit, though inexpensive and convenient for constructing Geniacs, is however not necessary; and some persons will prefer to construct their Geniacs using other materials."

This was the crowning lesson: It was the *form* of the machine, not its substance, that mattered. If you wanted to really strip the idea of a computer to its essence, you didn't necessarily have to make it electrical. One could even be constructed from Tinkertoys.

2 | Tinkertoy Logic

In the lobby of the Boston Science Museum, protected by a Plexiglas case, stands a whimsical monument to this most incisive of ideas: the power of a computer arises not from the nature of its parts—be they transistors, microchips, vacuum tubes, wooden wheels, or pulleys and string—but from its architecture, how the parts are arranged. Flip open the side door of, say, a Macintosh G4 and the electronic innards look like a space-age city viewed from the sky. Copper-colored avenues shuttle electronic traffic between squat black structures reminiscent of suburban office buildings: the computer chips that, if magnified many times, look like little cities in themselves, populated by billions of microscopic structures laced together by wires whose thickness is measured in millionths of an inch. This image has become so familiar that people now habitually reverse the metaphor, describing a suburban view from an airplane as looking like the layout of a chip.

This cartographic complexity seems to be the quintessence of computing. But for all the apparent differences, there is nothing going on inside the mightiest Macintosh or Dell or Compaq—or for that matter in the supercomputers like Blue Mountain and Q—that cannot be done, in principle, with a stadium full of Geniacs, or even a sprawling contraption made from Tinkertoys, like the one in the Boston museum. In the mid-1970s, a group of computer-science students (including Daniel Hillis, who went on to start a company called Thinking Machines) gathered together more than a hundred Giant Engineer Tinkertoy sets, assembling the thousands of wooden spokes and spools into a computer that played a mean game of tic-tac-toe.

Figure 2.1. Tinkertoys: an advertisement from the early twentieth century

Anyone who cracks open the most basic high-school text on computer theory almost immediately learns two comfortingly simple truths: Any kind of data, from moves on a checkerboard to streaming videos, can be translated into a code consisting of just two symbols, 1 and 0. And the data can be manipulated using basic operations called AND, OR, and NOT, trivial functions that can be carried out with simple little switches called gates.

If an "AND gate" receives a 1 signal at both of its inputs, A *and* B, then the device responds with a 1: Yes. Otherwise it says

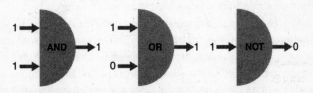

Figure 2.2. AND, OR, and NOT: the atoms of computation

No: 0. An OR gate is less restrictive: it will fire if it receives a 1 at either A *or* B. And a NOT gate does nothing more than invert its input: If it receives a 1 it will say 0, if it receives a 0 it will say 1. Weave together millions of these elements and the result is a digital chain reaction. Ones and zeroes, representing words, numbers, pictures, sounds, or positions on a game board, ricochet through the circuitry like pachinko balls. A computation is performed.

It's easy to see how such simple things could be made from ordinary switches and wire, as they were with the Geniac. In an AND gate, two switches are connected one after the other so that both must be closed for current to flow and the bulb to light. In an OR gate, the switches are in parallel; closing either one will complete the circuit. But it is merely incidental that the gates just described are electrical. The students, studying at the Massachusetts Institute of Technology, took on the challenge of making them from the sticks and spoked wheels in a child's set of Tinkertoys.

Suppose you have a Tinkertoy OR gate (as shown below). If either of the input shafts, A *or* B, is pushed to the right, the output shaft will move too. NOT and AND gates can also be constructed with Tinkertoys. By combining hundreds of gates, the

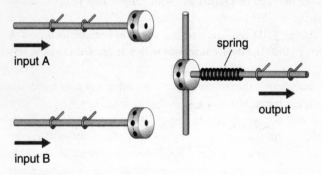

Figure 2.3. An OR gate made from Tinkertoys

MIT students built a complex mechanical contraption that would respond to any of its opponent's moves. Clicking and clacking, the wooden pieces, each positioned just so, would process the information, finally coming back with the best countermove.

Tic-tac-toe is particularly easy to code into the all-or-nothing diction of the binary tongue. Every round of the game leaves the board in a different state, with each of the nine cells of the criss-cross pattern containing an X or an O or remaining unoccupied. After studying the game, the students made a list of the best way for a player to react to every situation. Suppose the human opponent is playing O and the board looks like this:

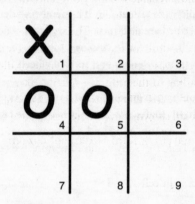

Figure 2.4. Tic-tac-toe

Obviously the computer can stave off defeat only by putting its X in the right-hand position of the middle row.

Without worrying about the details, one can imagine how this situation might be captured by a combination of rackety Tinkertoy gates manipulating Xs and Os instead of 1s and 0s. First, number the cells of the board starting in the upper-left corner. If cell 4 AND cell 5 are marked O, then put an X in cell 6. One can almost hear the wooden shafts sliding into place.

Referring to their list of moves and countermoves, the students built Tinkertoy "circuits" for each one. Linking AND and OR gates, so that the output stick on one would move an input stick on another, required additional parts, like string, which was held taut with fishing sinkers. This turned out to be the machine's weakest link. The string would stretch, throwing off the calculations. A part that should be in one position for X and another position for O might get stuck in between. But the logic behind the machine's design was impeccable. For every game it either won, tied, or (as often as not) malfunctioned. But in a platonic world of pure idea it would have been unbeatable.

There is more than one way to get Tinkertoys to play tic-tac-toe. A later, more reliable version (long since dismantled) took an entirely different approach. Tinkertoys were ingeniously assembled into the physical instantiation of a lookup table that gave the best response for each move. In this new system, each of the nine cells was represented by a triplet of digits: a 1 and two 0s. According to the arbitrary system, 100 meant that the cell was marked X, 010 that it was marked O, and 001 that it was blank. The board shown above would be encoded like this:

100 001 001 010 010 001 001 001 001
X   O   O

There is an X in cell 1, and Os in cells 4 and 5. The rest are empty.

In a conventional computer each of these number strings could be stored as a sequence of magnetic smidgens on a hard disk. The presence of a spot would be 1, its absence 0. Instead, the MIT students used a shaft, or "memory spindle," made from Tinkertoy spools and sticks. The presence or absence of a spool in a certain location indicated whether a cell was blank or filled with an X or O. With a little ingenuity the students were able to capture all the relevant board configurations using just 48 spindles—a wooden database.

With this knowledge in place, the machine was ready to play.

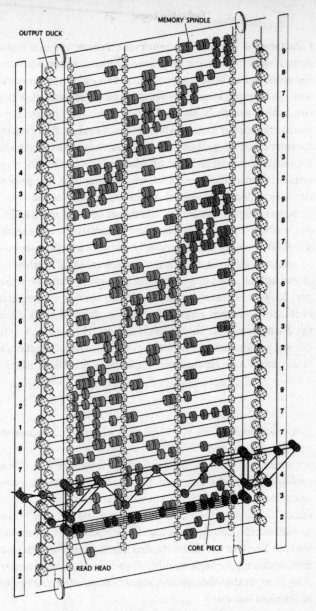

Figure 2.5. A tic-tac-toe database (Hank Iken for *Scientific American*)

Each time the human opponent made a move, the new board pattern would be entered into the machine by properly arranging spools on an "input stick." Attached to each of these spools was a protruding finger. The result was a mechanical claw that combed its way through the Tinkertoy database feeling for a match. When the pattern of Tinkertoy spools on one of the memory spindles meshed with the pattern on the input stick, the latter would rotate. This action caused an output contraption, called a "duck," to swing down, pecking at the number of the cell in which the computer wished to make its mark.

Both Tinkertoy computers, the earlier version and the later one, embody the same abstract idea: Information can be represented by anything that can reside in one of two distinct states—a switch turned on or off or a Tinkertoy that is here or there.

Instead of tic-tac-toe moves, the data to manipulate could consist of numbers to add and subtract. The trick is learning to count using just two fingers instead of ten, representing everything with 1s and 0s, the binary digits called bits.

Binary arithmetic can take some getting used to. Think of the odometer of a car. It consists of a row of wheels each marked with ten symbols, 0 through 9. When the wheel on the far right has completed a single revolution, ticking off 0 through 9, the wheel to its left moves up a notch to 1. It is registering the fact that it has counted one ten. The next click is 11. It has counted one ten and one one. Then 12, one ten and two ones. When the ten wheel has gone all the way around, counting all ten tens, the "hundreds" wheel registers a 1 and the cycle begins anew. The number 227 is made of two hundreds, two tens, and seven ones.

The number of digits on the wheel is arbitrary. One could also use 3 or 8 or 16. When the first wheel completed a revolution, advancing through 16 clicks, the next wheel would register a 1. (Thus, in this unfamiliar system, "11" would mean 17: one sixteen and one one.)

The simplest kind of counting machine would have just two

Figure 2.6. A base-10 odometer like one on the dashboard of a car

hundreds   tens   ones

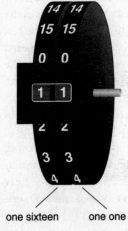

one sixteen   one one

Figure 2.7. A base-16 odometer with six more digits on each wheel

| 1 | 1 | 0 | 1 | = 13 |
| eights | fours | twos | ones | |

Figure 2.8. A base-2, or binary, odometer. Each "wheel" (more like a playing card) has just two digits, 1 and 0.

numbers, 0 and 1. In this system, 0 is still 0 and 1 is still 1, but then the supply of symbols runs out. Once the first wheel has ticked through both its digits, the next wheel registers 1. So "10" (one two and zero ones) means 2 and "11" (one two and one one) means 3.

Starting the cycle again, 4 is 100: one four, zero twos, and zero ones. What we ten-fingered primates call 5 is 101: one four, zero twos, and one one. Continuing on, 6 is represented as 110 and 7 by 111. And in the next cycle comes 8 (1000), 9 (1001), 10 (1010), 11 (1011), 12 (1100), 13 (1101), 14 (1110), and 15 (1111). Reading from right to left, 13 consists of one one, zero twos, one four, and one eight. Here it is in Tinkertoys:

Figure 2.9. The number 13 rendered in Tinkertoys

Any number could be encoded with sticks and spools and, with considerable mechanical skill, manipulated with AND, OR, and NOT gates in a way that resulted in addition, subtraction, multiplication, or division. It's not important to see how actually to design the Rube Goldberg circuitry, but to savor the simplicity of the idea.

Imagine a Tinkertoy spell checker, in which each letter of the alphabet is assigned a different number—a unique pattern of 1s and 0s, spools arranged on a shaft. One could design a machine that checked words, in the form of Tinkertoy constructions, against a vast wooden database (made from memory spindles) encoded with the proper spellings. Given an unlimited supply of Tinkertoys, an entire text could be rendered this way. Words would be inserted or deleted by moving around Tinkertoys. Like letters of the alphabet, the range of musical tones or colors can also be assigned numbers and translated into clusters of bits. Imagine Tinkertoy spindles, light-years long, encoding motion pictures.

Making something so huge and unwieldy is possible only in the mind (or in a computer simulation). But the lesson stands: Any computer no matter how complex consists of a bunch of little objects that can be in two positions, 1 or 0. Allow them to interact and any pattern—a mathematical equation, a novel, a symphony, a painting, or a movie—can be stored and processed.

The beauty of the binary code lies in this irreducible simplicity. One could also make a Tinkertoy machine, or any computer, using the more familiar "base-10" number system. Then every spool, or whatever is used to encode information, would be capable of assuming 10 different positions. But once you require that level of acuity, the danger for error grows, mistaking a 2 for a 3 or an 8 for a 9. How much simpler to worry only about whether something is present or absent, here or there, 1 or 0. If a bit is slightly off, 0.021 instead of 0 or 1.043 instead of 1, it can be automatically snapped back into the closest allowable position. (It was the lack of this ability, called "restoring logic," that plagued the first Tinkertoy machine. It needed a way, using springs perhaps, or rubber bands, to keep spools from drifting into intermediate limbo.)

With binary magic, the digital recording of a song, indicated by the presence or absence of microscopic pits on a compact disc, can be read by the luminescent finger of a CD player's laser beam and then converted into an electrical signal—high voltage is 1, low voltage is 0. And that data can be transformed by a computer and modem into the high and low squeals that carry digits through telephone wires, allowing the music to be "uploaded" onto another computer on the Internet, stored now as the presence or absence of magnetic specks on a disk drive. From there it can be downloaded and uploaded again and again. As long as 1s are recognized as 1s and 0s as 0s, with nothing in between, the copy and the copies of the copies are as faithful as the original.

After all, the hallmark of a digital computer is that it is deterministic. Every cause leads to a predictable effect. Send a string of bits—numbers to be added, words to be sorted—into the circuitry and the bits will rattle through the gates until out the other end comes the answer. Every time the process is repeated the result should be the same. That, of course, is the whole point of a device in which every component can be only one way or the other with no ambiguity allowed.

Even the improved version of the Tinkertoy computer never operated very reliably in the real world. Bits can be manipulated more fluidly if they consist of electrons instead of wooden spools. And far more circuitry can be packed into a smaller space. In early computers, bits were batted about with clacking electrical relays, mechanical switches that could be turned on and off with the tug of an electromagnet. The arms of the relays were used to open and close circuits that powered other electromagnets, sending swarms of bits cascading through the machine. Relays were soon replaced by vacuum tubes.

Today's switches are silent and thousands of times faster—

tiny silicon transistors that can be in two states: on, allowing electrons to pass through them, or off, interrupting the flow. At first a single transistor was enclosed in a metal case about the size of a pencil eraser with three wires protruding from one end. Electricity would enter through one wire and exit from the other, and a signal applied to the middle wire could be used to switch the current on and off.

Dozens of these separate units were soldered together to make electronic circuits. Then engineers discovered how to etch several transistors onto the same wafer—ten then a hundred then a thousand. Before long, millions of transistors could be squeezed onto a single chip along with miles of microscopic wiring. Every year, as technological nimbleness increases, so does the density of the circuitry. The advantage is not just smaller components. As the distance between the little switches contracts, information can flow faster, and so computing speed grows.

Along the way the machines have become increasingly more flexible—programmable. The Tinkertoy computer is a dedicated machine, its parts carefully arranged to execute a single task. To make the clacking sticks and spools do something else, the designers would have to take them apart and rearrange the pieces. In the early electrical computers, the relays could be reconfigured to perform a variety of tasks by plugging cables into different sockets like those on an old-fashioned telephone switchboard. The instructions for where to put the plugs, written out in a loose-leaf binder or a spiral notebook, were the first computer programs, software in its crudest form.

The next step came when engineers realized that the plugs didn't have to be moved around by hand. Paper cards or tapes punched with patterns of holes could be fed into the machine, opening and closing circuits automatically. The bits streaming in represented not just the data to be processed but the instructions for how to process them—how to connect all the little switches in a way that would carry out the proper mission.

With a modern PC we blithely double-click an icon on the desktop, summoning a flow of data from the disk drive—the pattern of bits that configures thousands of little switches to act as a word processor or a Web browser or an MP3 player—temporary little structures, virtual machines. They are allowed to exist only as long as they are needed. Then they are wiped away and replaced with other structures, all built from 1s and 0s.

It is hard to believe sometimes how well this all works. You can call up a movie trailer in a window and drag the animated image around the desktop, causing millions of bits to pour through the computer's hidden registers. It is overwhelming to try and imagine the precise coordination going on behind the screen. Ultimately, though, it all comes down to shuffling 1s and 0s, flipping little switches on and off. Tiny invisible Tinkertoys.

3 | Playing with Mirrors

A well-known adage called Moore's Law predicts that the number of components on a chip doubles about every two years. The Pentium 4 processor introduced at the beginning of the millennium is packed with millions of little switches silently clattering more than a billion times a second. But there is a lot of shrinking yet to be done. Small as they are, the switches are each made from a billion or so atoms. If Moore's Law continues to hold, the logical outcome is a switch a single atom in size.

A chip filled with so many submicroscopic devices would be formidable indeed. But something more fundamental happens at this scale. Quantum mechanics kicks in. Anyone who watches *Star Trek* has at least a dim notion of what that means: particles that can be in two or more places at once, that can seem at one moment like hard little specks of matter and the next like waves.

What if atoms with their surrealistic behavior were used as tokens to compute? Spinning counterclockwise, with its axis of rotation pointing upward like a top, an atom could be decreed to represent the digit 1. Flip it over so it is pointing upside down (and spinning clockwise) and it can represent the digit 0. (The designations of course are arbitrary; the labels could just as easily be reversed.) If those were the only possibilities, then an atom would just be an unimaginably small version of an ordinary switch. Quantum mechanics tells us, however, that an atom can also be in both states, 1 and 0, at the same time.

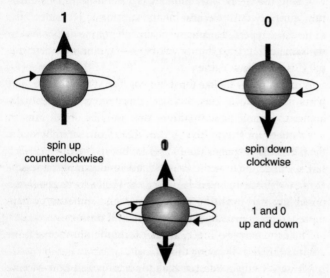

Figure 3.1. Spinning atoms. Think of them as little tops.

It is fruitless to try to understand this in the usual manner, by making little pictures in your head. It is as if each spool in the Tinkertoy computer could be not just here or there but simultaneously here and there, something like a double-exposed photograph. But that is not right either. Quantum theory is not a matter of darkroom trickery and imaginary worlds. The atom really is in both states at once. Impossible but true. In the decades since quantum mechanics was discovered, physicists and philosophers have not stopped arguing over what it means. They will probably be debating it a century from now. It's just something you have to get used to. Accept it and move along.

When the computer engineers' incredible shrinking act reaches the quantum level, the world inside a chip will no longer be deterministic, the 1s and 0s unambiguously distinct. There will be 1s, 0s, and Φs. (This symbol is used here not to represent the Greek letter phi, but as a stand-in for this quantum curiosity 0/1.) Because of the subatomic blurriness, the same cause repeated again and again will not inevitably lead to the same effect. Uncertainty will reign—"quantum indeterminacy" is the official name.

At first this seems like terrible news. For years, as computer parts have grown ever smaller, threatening to eventually approach atomic size, engineers have worried about how to *avoid* quantum effects, how to keep them from scrambling the data, mashing together the 0s and 1s. But in the early 1980s a few scientists such as the late Richard Feynman and a lesser-known visionary named Paul Benioff began to consider the possibility that quantum uncertainty could someday be harnessed to make machines of unprecedented power.

The first chapter of the computer revolution has come from automating and shrinking digital logic, speeding it and speeding it until Geniacs become Blue Mountains and Blue Mountains become Qs. What might turn out to be part 2 of the story springs from this new, counterintuitive approach: tapping into nature's own peculiar logic, igniting quantum fires. The old notions of computation begin to crumble, and we're no longer playing with Tinkertoys.

I confess that, on the deepest level, I find quantum mechanics hard to swallow. I know I'm not alone. The brain was molded by evolution to guide people through a world of objects made from many atoms, so many that the weird quantum effects are not evident. Things appear to be either here or there, spinning this way or that. Whatever progress can be made in grasping how opposites can coexist on the atomic realm comes from acknowledging the narrowness of the human point of view. The universe doesn't care whether we believe in its rules. When it comes to atoms and smaller things like electrons and protons, what is intuitively obvious—that something must be in one state or another—turns out to be the prejudice of creatures who haven't had to deal regularly with the extremely small. We are blinded by macrocentrism.

Physicists aren't much help. One of their favorite pastimes is lamenting the crazy notions the public has absorbed about the quantum world. They look down on books like *The Dancing Wu-Li Masters* and *The Tao of Physics,* which suggest that the ambiguous nature of quantum matter is connected with the duality of Eastern religion—yin and yang and all that. There is little reason to suppose that the precise truths of quantum theory can be divined through meditation or anything other than the hard route physics took: observing how nature behaves and devising a theory—the only possible theory, it seems—that makes sense of it. But there is no getting around the fact that quantum theory is very weird.

Its founder, Max Planck, never imagined the implications he was about to unleash when he found, in 1900, that light seems to be emitted in little packets or particles—the quanta that give the theory its name. His is not an easy argument to follow, but here is the gist: In geometry class we learn that a line is a continuum, meaning that no matter how short it is, there are an infinite number of points between its two ends. You can cut the line in half and then cut the half in half, and then halve that half . . . ad infinitum. It is made of an infinite number of infinitesimally tiny points. Planck showed mathematically that if, as everyone had long assumed, energy is also smooth and continuous— made from an infinite number of infinitely tiny portions—then a hot, glowing object should radiate an infinite amount of light, an obviously absurd prediction. But if one assumed that the energy came in discrete little packets, or quanta, with a minimum possible size, the paradox went away.

Einstein clinched the case a few years later, marshaling quantum theory to explain a phenomenon called the photoelectric effect: Light shining on certain metals causes them to spit out electrons. And this, Einstein showed, could be explained if light came in lumps.

Experiment after experiment has corroborated the idea. Feynman once put it like this: It takes five or six light quanta, or photons, to tickle the human retina into firing off a signal to the brain. If our eyes were a bit more sensitive, we would see each individual particle. Light would appear to come at us in pulses, like bursts from a Roman candle. *Light comes in packets.* The quantum weirdness that has become such a staple of popular science and science fiction flows from this seemingly innocent observation.

Though we can't see the pulses, they manifest themselves in the everyday world. Look through a window at the landscape outside, or whatever is on the other side of the glass. Hovering in the background, you may see a ghostly image of your own reflection. Light particles, bouncing off your body, hit the glass. While some of these photons come bouncing back and into your eyes, most of them sail on through. (Someone on the other side of the window can see you, along with a dim reflection of his own image.)

What determines whether each individual photon, reaching the glass, takes one route or the other? It is natural to think that there must be something different about photons that pass through the window and photons that come reflecting back. But all photons are, by definition, the same. Do the photons hit the glass at different speeds or with varying amounts of wobble or spin? Are they affected along the way by disturbances in the atmosphere? All those possibilities have been experimentally ruled out. No matter how stringently one controls the variables, the perplexing effect remains: As far as can be determined, two photons traveling under identical conditions, fired one after the

other, can hit the glass in the very same spot, and one will ricochet while the other one flies through (or both may ricochet or
both may not). It took physicists years to get used to what is
now considered all but incontrovertible: Each photon makes
the "decision" at random.

This is not the familiar kind of randomness that comes from
living in an imperfect world. If a billiard ball hits the edge of a
table at a 45-degree angle, it will bounce 45 degrees in the other
direction—in theory. Tiny imperfections in the ball and the
table ensure that there will always be slight departures from the
ideal. But with finer craftsmanship, the imperfections can be
reduced as much as we like. The ball can be made more precisely spherical, the table more precisely flat.

If billiards is played with photons and mirrors, no amount of
skilled honing can ever completely eliminate the uncertainty.
There is a threshold below which you cannot go. If a stream of
photons traveling precisely at 45 degrees strikes the perfectly
flat surface of the mirror, most of them will bounce off at 45
degrees. But the rest will ricochet at a variety of different angles,
and there is nothing anyone can do about it.

This randomness arises not from human ignorance or from
the crudeness of our perceptions—not even from the clumsiness of our experimental techniques. Take all of that into
account and the randomness remains. Quantum theory tells us
that we can speak with assurance only about what, *on average,*
whole hordes of photons will do. There is no way to predict the
behavior of a solitary photon. The best one can do is to talk in
probabilities: there is, say, a 99 percent chance that the photon
will bounce at 45 degrees, and a whole range of lesser chances
that it will follow other routes. This inherent uncertainty is part
of the weave of the universe.

The great achievement of the inventors of quantum mechanics was to devise equations that describe this puzzling behavior.
And what the equations say is that the photon, traveling

through space toward a window, exists in a kind of limbo in which all the possible outcomes of its journey hover together simultaneously—a state called a "quantum superposition." "Pass through glass," "bounce straight back," "bounce back at 45 degrees," "bounce back at 30 degrees": all these choices cling together as the photon moves. When it hits the glass, the bundle of possibilities comes undone. One of the outcomes is randomly chosen. Again, the choice has nothing to do with slight imperfections not taken into account in the analysis. The randomness is fundamental.

There are no perfect analogies with the reality we perceive, no way to get a firm mental grip. But approximations can help. Think of the music coming from the vibrating cone of a loudspeaker. This undulating wave is made from a complex of interlaced sounds: "wavelets" that represent violins, cellos, violas, cymbals, timpani. The photon traveling toward the glass can be thought of as a different kind of wave, one that is woven from wavelets each describing one of the particle's many possible states—the likelihood that it will act a certain way. When the photon reaches the glass, this delicate juxtaposition is disturbed. One of the wavelets survives the crash: The others disappear.

The idea seems so absurd that it is tempting to dismiss these quantum waves as mathematical contrivances—a clever tool for making calculations about particles, not something of substance. But in the laboratory these "probability waves" can be manipulated with mirrors and filters, treated almost like waves of water or waves of air. Waves of stuff.

This philosophically jarring truth has become enshrined in physics as the two-slit experiment. Start with a pan of water. Across the middle put a barrier with two holes in it. Drop a small object in the left-hand pool and a wave will radiate toward the divide. When it reaches the holes, it will split into two waves, which pass into the right-hand pool and interact. If two crests overlap they reinforce each other, intensifying the wave. If a crest lines up with a trough, they cancel each other out. The result is called an "interference pattern"—the hallmark of wavelike motion.

Then comes act two of the demonstration. Shine a beam of photons at a piece of paper, and between the source and its target place the barrier with two holes. First cover one of the openings so that a single blob of light is projected onto the paper. Then cover the second hole instead. The blob shifts from one position to the other. Finally open both holes. The image on the other side is not, as might be expected, two blobs of light side by side. The waves coming from both slits interfere with each other like water waves and leave a pattern of light and dark stripes, marking the places where the waves reinforce or offset each other.

Figure 3.2. The two-slit experiment. The wave goes through two holes in the barrier. On the other side, the two resulting wavelets interfere with each other.

Light, then, is obviously a wave. If that were all there was to the experiment, it might seem that light is simply a wave of substance, made of photons instead of water molecules, not so mysterious after all. When it reaches the barrier, the wave splits and the two smaller waves interfere with each other like the waves in the pool. What is so weird about that?

The answer comes when you turn down the intensity of the beam, slowing the stream of photons, until only one particle at a time pulses toward the barrier. (To detect where each particle lands, the target can be replaced with an array of photosensitive detectors or a piece of photographic film.) Each single photon surely must go through one hole or the other. With nothing to interfere with the journey the wavelike effects should disappear. The result, as the impacts accumulate, should be two blobs of light on the screen with no interference effect. What, after all, would interfere with what? But that is not what happens. Point by point, the little winks of light trace out the same striped interference pattern. It's almost as though an onboard computer were telling each photon where to go.

What the experiment really seems to demonstrate is that light isn't a wave of many photons in the way that rippling water is a wave of H_2O. Each photon is *itself* a wave, a superposition of its numerous possible states. It is this wave of possibilities that hits the barrier, splitting into two smaller waves that interact. Some of the possibilities are strengthened while others weaken or disappear. The combined wave that emerges then strikes the target, its superposition collapses, and just one of the likelihoods is realized—as a tiny spot on the film. What seems like a mathematical abstraction is acting like a real thing. The next photon will probably strike in a different place, and the next one in yet another, and over time the light and dark stripes emerge.

Trying to make sense of this, the mind struggles with imperfect metaphors. Feynman preferred to think of the photon not as a probability wave but as a single particle with extraordinary properties, able to simultaneously travel through both slits, or to sail through a window while also bouncing back. He called this the "many-paths" interpretation of quantum mechanics.

Think of a light beam bouncing off the center of a mirror. Made from many photons traveling en masse, the beam takes the most efficient route, with the angle at which it approaches the mirror equaling the angle at which it reflects. But what is

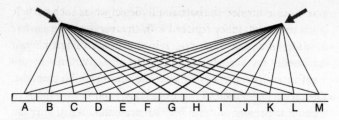

Figure 3.3. A light beam bouncing off a mirror. According to Feynman's "many-paths" interpretation, the photon can be thought of as trying out every possible path simultaneously.

happening behind the scenes? The beam is made from quantum particles, so they must each behave probabilistically. A single photon is most likely to aim for the middle of the mirror, taking the most direct route. But it is free to follow any trajectory at all. It might bounce a little to the left or right of center, or—less likely, but possible—fly over to a spot on the far edge of the mirror. In fact, according to the "many-paths" perspective, each single photon is trying out every possible path simultaneously. The trajectories interfere with one another, the likelier ones adding together, the least likely canceling out—finally yielding the expected outcome: the familiar path in which the beam bounces off the center of the mirror, approaching and retreating at equal angles.

It is a matter of taste which is stranger: this, or the wave analogy in which the photon is an undulating superposition of possibilities that collapses into a single outcome. Over the years physicists and philosophers have produced a whole library of rival interpretations. (In the "many-worlds" version of quantum mechanics, one imagines that each possibility actually occurs in a separate parallel universe—the source of an endless supply of science fiction scripts.) All the different perspectives have been shown to be mathematically equivalent and all come down to this: Human brains are just not equipped to intuitively understand the subatomic rules. There is no reason why they should be. In the world people evolved in, only a few physicists deal with individual particles. More commonly, the particles come at us en masse. With their constant interactions, they disturb one another, ensuring that the mysterious probability waves almost immediately collapse. Quantum behavior is masked, so understanding it has not been necessary for survival.

What is true for photons is true for all particles: the electrons that hover in the shells of atoms and flow through wires; the protons and neutrons that make up the atomic cores. At any one moment, a particle is not in a precise place but in a superposition of all its possible locations. Nor does this quantum ambiguity apply only to position. A particle can also be in a superposition in which it spins clockwise and counterclockwise, down and up. Or a wave packet can represent all of a particle's possible velocities. Only when it is measured or observed—subjected to some kind of disturbance from the world outside—does the probability wave randomly collapse. The multitude of might-have-beens narrows into a single actuality: what we mean when we say something is "real."

The fundamental truth of quantum indeterminacy has become so deeply ingrained that anyone who seeks to overturn it immediately faces the obstacle of proving he is not a crank. Even Einstein's doubts about the theory are now seen by admirers as a rare occasion when he succumbed to a prejudice about how the world should really be. In 1935, driven by his famous dictum that God does not play dice with the universe, he and two young colleagues, Boris Podolsky and Nathan Rosen, tried to prove that quantum mechanics is mistaken, that an unobserved particle moving through space has the same hard-edged attributes as any physical object. If it seems to be in a fuzzy state of uncertainty, randomly snapping into focus only when it is measured, that is just a sign of human ignorance. Something must be missing from the theory.

Einstein liked to challenge commonplace assumptions with fanciful thought experiments. An early breakthrough in his theory of relativity came when he considered the contradictions that would arise if one could travel fast enough to catch up with a beam of light. The waves would appear to be standing still, it seemed, and it wouldn't be light anymore.

In their attack on quantum mechanics, he and his collaborators tried to prove that it led to an absurd conclusion. In what has become known as the EPR (Einstein-Podolsky-Rosen) argument, they imagined a particle that decays to produce two photons flying off in opposite directions. Photons are endowed with the quality called spin and can be roughly thought of as rotating either clockwise or counterclockwise. Spin, like energy, is conserved: If you add up the amount before and after any particle interaction, the two quantities must match. The decaying particle in the EPR experiment has no spin, so the two photons it produces must be rotating in opposite directions, for a net effect of 0.

Quantum mechanics, however, tells us that until the photons are measured neither has definite attributes. They each exist in a superposition, spinning clockwise and counterclockwise at the same time. This is the idea Einstein and his colleagues found so annoying. Surely, they believed, each must be in a definite state from the moment it is spawned, turning one way or the other. The measurement simply confirms what was true all along.

To believe otherwise, the scientists proposed, would lead to a contradiction. Suppose that several moments after the decay, you measured one of the photons, forcing it to "choose" between one spin or the other. Whichever direction it takes, the other photon must immediately snap into the opposite state so that the total spin is 0. But no time has passed in which any kind of signal could travel from one particle to the other, allowing them to synchronize. Any communication would have to be instantaneous, faster than light. Since quantum mechanics leads to this absurdity, it must be mistaken or incomplete.

It is now considered firmly established that the absurdity is true. The photons cannot instantaneously exchange information, in violation of the laws of relativity. But they are indeed somehow linked—or "entangled," as physicists say—in a way that defies ordinary conceptions of locality, the notion that something is in one and only one place at a given time. Since Einstein's thought experiment, real demonstrations in laboratories have confirmed the EPR effect. Particles in different locations, no matter how distant, can be somehow correlated, their fates tied together in a way that involves no ordinary physical connection. If one is spinning up, the other must be spinning down. It is almost as though the two photons were not separate entities but two faces of some larger thing.

Physicists call the orderly, predictable world of billiard balls the classical realm, for it obeys the laws of physics as they were understood before quantum mechanics came along. For every action there is an equal and opposite reaction. For every cause there is a predictable effect. In practice, there are limits to the fineness with which ordinary events can be measured. But these imprecisions arise from human ignorance and clumsiness and, with effort, can be indefinitely improved upon.

The digital computer is firmly rooted in the classical world. Through great feats of engineering, uncertainty has been reduced to the point that the same series of 1s and 0s sent through a web of circuitry again and again will produce the same result: 1 + 1 will always equal 2, and if it doesn't we can tweak the system to make it so. Quantum uncertainty isn't a problem. Since a tiny switch on a chip is made from a galaxy of atoms—constantly interacting, nullifying the quantum effects—it follows the laws of classical physics. It is definitively on or off. We don't need to worry about the switch being in quantum limbo.

The result is an obediently deterministic contraption that can be used to simulate other deterministic contraptions: the mechanical motion inside a gasoline engine, the patterns of traffic in a city, or the workings of another computer. In fact, given enough memory and time, any computer can simulate any other computer. This is one of the deepest principles of computer science. An ancient Radio Shack or Apple with a cassette tape drive for a memory can mimic a supercomputer like Q. Churning through all the calculations might take centuries and warehouses full of tape, but for both machines the most basic operations, the shuffling of 1s and 0s, are equivalent. Computers are universal machines.

But try to use a digital computer to simulate a quantum system—electrons moving around a nucleus, subatomic particles colliding—and its powers are pushed beyond the limit. If the most fundamental thing about a digital computer is that it behaves in a precisely predictable manner, the most fundamental thing about a quantum system is that it does not. The result is a clash between two different worlds.

Say you want to simulate a group of 10 particles spinning either clockwise or counterclockwise. Each particle can be represented by a bit of information (a switch turned on or off, a Tinkertoy pushed to the left or the right): 0 can be decreed to mean clockwise and 1 to mean counterclockwise. We, the designers, are as free to attach the labels as we are with the Geniac.

Now count up all the possible states the particles might be in. All of them spinning clockwise would be represented by a row of 10 switches each set to 0. All spinning counterclockwise

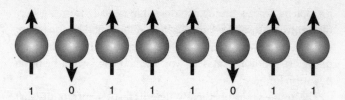

Figure 3.4. A row of atoms spinning clockwise (0) or counterclockwise (1)

would be represented by a row of solid 1s. And then there are all the cases in between. The first particle spinning counterclockwise, and the rest clockwise. The second particle spinning counterclockwise . . . or the third . . .

Calculating the total number of possible patterns is simple arithmetic. Ten particles that can each spin two different ways can be in $2 \times 2 \times 2 \times 2 \times 2 \times 2 \times 2 \times 2 \times 2 \times 2$ or 2^{10} different states, a total of 1,024. Each of these patterns would require a separate row of switches to represent it, 1,024 of them lined up side by side. The first would say 0000000000; the last would say 1111111111. (This, according to our binary odometer, means one one, one two, one four, one eight . . . or $1 + 2 + 4 + 8 + 16 + 32 + 64 + 128 + 256 + 512 = 1023$. Counting the beginning row of all 0s, that's a total of 1,024 numbers.)

Start adding particles to the system and the required computing resources explode exponentially. Eleven particles require 2^{11} or 2,048 rows of switches. Twelve particles require 4,096. Double the number of particles to 20 and you have not twice as many possible patterns but 2^{20} or 1,048,576. Forty particles require 2^{40}, or about a trillion rows of processors.

This much would be true as well if we were simulating classical objects, like marbles that are either black or white. Since we're dealing with quantum objects this is just the beginning of the complications, for each particle can also spin both clockwise and counterclockwise at the same time. (In fact, it can spin in every possible combination of the two directions: 99 percent

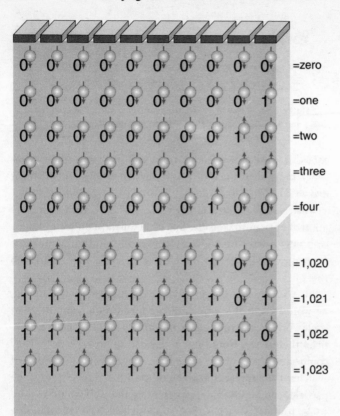

Figure 3.5a. A single row of 10 switches in an ordinary computer, each turned on or off, can represent any one number between 0 and 1,023 (1111111111 in binary). To represent them all we would need 1,024 rows of switches.

clockwise and 1 percent counterclockwise, 98 percent clockwise and 2 percent counterclockwise. . . .) Considered as a whole, the system is in an incredibly complex superposition of every way the particles might conceivably be arranged—an undulating probability wave. And that is only a snapshot of a single

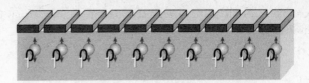

Figure 3.5b. A row of 10 quantum switches, which can each say 1 and 0 simultaneously, can hold all 1,024 numbers at once.

moment. From there the particles will interact and the computer must keep track of every way the whole system might evolve.

The result is an intractable problem. There is such a proliferation of options to consider that even the most powerful supercomputer would immediately choke on the data. A machine from the classical realm is straining to model a phenomenon that obeys very different rules.

In 1982, Feynman suggested a solution. What would happen, he wondered, if the simulation were run on a computer that itself worked quantum mechanically? Its switches, like the particles themselves, would not be limited to saying 1 or 0. They could say 1 and 0 at the same time. Since a row of 10 classical switches can represent just one of the 1,024 different ways the particles can be spinning, representing them all would require 1,024 rows of switches. But suppose you have 10 *quantum* switches, one for each particle. Since they can each be in both states at once, a single row can represent all 1,024 possibilities simultaneously.

This is not an easy idea to get one's mind around, and for now it is tempting to just take it on faith: a row consisting of x atoms (or other quantum particles) can hold 2^x numbers at the same time. Who would have imagined such a thing could be true? But the notion is so important to understanding quantum computing that it is worth taking a closer look. And to simplify, focus on just three quantum switches. There are eight possible three-bit patterns representing the numbers from 0 through 7.

000	0
001	1
010	2
011	3
100	4
101	5
110	6
111	7

Suppose that only the right-hand switch in the trio is in a superposition of 1 and 0. If the first two switches are both set to 0, then the numbers 000 and 001 can be represented at the same time. If the middle digit is also in superposition, then 000, 001, 010, and 011 can all be accounted for. Put the left-hand switch in a quantum state as well and all eight patterns can be registered with just three atoms.

Start adding to the row of quantum bits and the possibilities soar. A single row of 40 switches could simultaneously represent every one of the trillion ways that 40 particles can be spinning. One quantum system is being used to simulate another. The exponential explosion that would bog down a classical computer is harnessed and put to work.

Simulating particles this way would be a powerful tool for physicists. In the few years after Feynman suggested his solution, even more exciting possibilities emerged. Maybe other problems that exploded exponentially on an ordinary computer, rendering them unsolvable, could be tamed by a quantum machine. A row of quantum bits—what have come to be called "qubits"—might be used to represent numbers, words, sounds, images, anything that can be encoded as a string of 1s and 0s. And with the powerful magic of quantum mechanics, a huge number of bit strings can be stored simultaneously and processed at the same time. The desirability of billions of potential chess moves might be weighed all at once, in quantum superposition. By limiting themselves to classical physics, computer scientists have barely scratched the surface. Waiting underneath may be a whole new way to compute.

4 | A Shortcut Through Time

The "desktop" of a modern computer spins an arresting cybernetic illusion. Extending from one edge of the screen to the other is a soothing visual expanse—a mottled marble surface, a wood-grained veneer, or perhaps something more idyllic, like a flowing field of wildflowers or the image of a mountain stream. Scattered across the decorative vista is a clutter of virtual stuff: fingernail-size pictures representing texts, sound recordings, video clips, folders, and the various tools at the user's disposal—a simulated typewriter, calculator, or spreadsheet, an e-mail program and browser for exploring the labyrinth of the World Wide Web.

Click twice on a word-processing document and it blossoms into a virtual piece of paper filled with words and sentences that can be fluidly rearranged. Double-click a sound file called "Rhapsody in Blue" and what looks like a tape deck appears, complete with volume controls, fast forward and rewind buttons—music to work by. And if you want a different kind of stimulation, pull down the "bookmarks" menu of the Web browser and jump to a live twenty-four-hour-a-day streaming video camera overlooking Times Square. A window opens onto a frenzied world of taxis, pedestrians, and pigeons fluttering by.

Ultimately everything you see is being painted one tiny dot at a time. Dozens of times a second the screen is swept by a stream of bits, a constantly changing sequence of binary data telling each of the million pixels whether, at any moment, it should be red, green, or blue, colors that can be melded to make any other. This technological feat is impressive enough in a television set, where there is a single coruscating picture to project. But in a computer, you can reach into the pointillist landscape with your mouse and alter the objects or move them around. When you press a simulated button or drag a file inside a folder, the folder into another folder, and the whole thing to the trash, the visual metaphor is so beguiling you forget that all you are really doing is using the motion of the mouse to issue a continuous sequence of commands—instructions for the computer to rearrange bits in its memory, patterns that correspond to the mirror world of pixels on the screen.

Sustaining this hallucination requires a staggering amount of information processing. The Times Square video cam captures image after image of 45th Street and Broadway, sending its signal bit by bit through the Internet. As the data reach your computer through a high-bandwidth line, they must be decoded and properly dispatched so that they animate the correct block of the screen. At the same time, the digital notes of "Rhapsody in Blue" might be flowing off the hard disk to be converted into sound and channeled to the loudspeakers. At any time you can adjust the volume, or drag the sound player or the Times Square window to a different part of the desktop—and the computer obediently tracks your mouse strokes and takes note of which icons you're uncovering and which you are now covering up. If a message arrives during the commotion, its presence might be signaled with a beep, a piano chord, or the flashing image of an envelope in the corner of the screen.

The greatest fiction being spun before the eyes is that all this activity is happening simultaneously. The conjuror of this illusion is a single computer chip. Deep inside the maze of hardware, all the competing flows of data are ultimately funneled through this master processor—a Pentium or a PowerPC—one small chunk at a time. In a typical personal computer, hundreds of millions of tiny calculations are performed by the chip every second. But for all its power, it is focusing myopically on a single task at a time. At one tick of the clock, it processes a small cluster of bits describing a pixel arriving from the Times Square cam, in the next instant it turns to a split-second sample of "Rhapsody in Blue," then to a chunk representing the letter or numeral you just typed.

Computer engineers call this the Von Neumann bottleneck, after the physicist John von Neumann, who helped design one of the earliest computers. In a modern processor a few micro-operations might be performed simultaneously. But the essence of the original architecture remains. The increasing computer power the world has become addicted to arises not so much from widening the neck of the digital funnel but from dramatically increasing the velocity of the flow.

In fact, strip away the fancy case and other incidentals and every computer is no more than a trumped-up version of a contraption called a Turing machine. In the 1930s, long before the first digital computers were made, the British mathematician Alan Turing invented an imaginary problem-solver consisting of just two parts: a clocklike dial with a pointer and a scanning head that can move alongside a lengthy paper tape examining, one by one, the marks it finds. A problem to be solved is coded onto the tape in a language consisting of two symbols, X and O, and the pointer is set to a specified starting position. Then the machine (think of it as a black box full of gears or electronic components) processes the data according to a list of simple instructions: "If there is an X in the square you are reading, and the clock dial is on position 5, replace the mark with a O. Then move the tape two spaces to the left, and reset the dial to position 2." The machine then reads the new mark in front of its nose and, referring to the rule table, follows the relevant command.

Figure 4.1. A Turing machine

With the right table of instructions (what is now called a program) the machine can take on any problem. If it is programmed to multiply and then given 2 and 3, coded perhaps as XXOXXX, it would crunch through the manipulations, one step at a time, and print XXXXXX. If programmed to raise the first number to the power of the second number, it would shuffle the tape back and forth, reading, printing, erasing, and finally leaving behind a trail of eight Xs. Computation, it seems, is simply a matter of taking an input string and converting it, according to the rules of the program, into an output string—an idea that will become important to understanding how a row of atoms computes.

More complex problems require longer lists of instructions, and if necessary the Turing machine can use part of the tape as a scratchpad—a memory—temporarily storing and then erasing intermediate results. Turing proved that, supplied with a limitless supply of tape, his machine could find the answer to any solvable problem. (If the problem was unsolvable, the machine would churn and churn forever.) Most presciently of all, he realized that the list of instructions for how to add, subtract, multiply, divide, take a square root, and so on did not have to be permanently wired into the machine. They too could be expressed in a language of Xs and Os and fed in along with the data.

The fundamental dogma of computer science is that nothing is more powerful than a Turing machine. You can make computers that are faster. You can put ten or a thousand or a million machines side by side all crunching away at a problem. (This kind of "parallel" computation is what supercomputers like Blue Mountain do.) But all you are doing is making more efficient Turing machines, glorified versions of the dexterous old tape scanner. No device that can be realized in the physical world can do anything that can't be done, given enough tape and time, by Turing's paradigmatic computer. It has only been in the last few years that this principle has had to be amended: No device *operating according to classical physics* can do better than a Turing machine.

Anyone struggling through a tedious calculation can identify with Turing's myopic bean counter. Suppose you want to find the factors of 245, two or more smaller numbers that can be multiplied to get the larger one. The most obvious procedure, or "algorithm," is a systematic version of trial and error. To factor 12, you start with 1×12, then 2×6, then 3×4. Five is not a factor, and 6 has already been accounted for. There is no point in going higher. Nothing greater than 6 can possibly divide evenly into 12 (except 12 itself). You've discovered a shortcut: it is only necessary to try each number up to one-half of the whole number. (And further scrutiny would reveal that you really only need to test numbers up to the square root.) If you try the procedure with 13, you quickly confirm that it has no factors other than itself and 1. It is a prime number, one of the irreducible particles of the number system. Any number is either a prime or a composite, one that can be produced by multiplying primes.

Splitting smaller numbers into primes—arithmetical fission—is a trivial matter: 49 is made from 7×7, 50 is made from $2 \times 5 \times 5$. But as the numbers to be decomposed grow larger, the calculating time increases at a staggering rate, undergoing an exponential explosion. Numbers a few hundred digits in length appear to be unfactorable in any conceivable amount of time, though it is possible that you might be extraordinarily lucky and just happen to guess the factors right away. But the slim chance that this will happen rapidly shrinks—exponentially implodes—for larger and larger numbers. (Since, for reasons we shall later see, some of the strongest codes on the Internet are based on factoring, interest in the problem extends far beyond the airy domain of number theorists.)

No one has absolutely ruled out the possibility that, lurking somewhere in the platonic heavens, is a super algorithm with the power to make factoring as fast and simple as arithmetic. It is possible that a bright thirteen-year-old playing with code on her iMac will stumble upon a solution. In *The Man Who Mistook His Wife for a Hat,* Oliver Sacks tells of a pair of autistic twins at a state mental hospital who entertained themselves by swapping large prime numbers, ones they had apparently generated in their heads. Sacks romantically attributes the curious phenomenon to a "Pythagorean sensibility," a mysterious neural resonance with the mathematical world. Maybe some accident in their neurological wiring gave them access to the very algorithm that has eluded mathematicians for so long. But there may be some less interesting explanation. As far as is known, breaking the back of the factoring problem would require something long thought impossible: a computer more powerful than a Turing machine.

The hardness of a problem can be measured by how long it takes to solve. But as computers grow faster, the old standards become obsolete. A modern pocket calculator can multiply two four-digit numbers as quickly as you can punch them in. Old mechanical calculating machines had to clank through the procedure one jarring step at a time.

It is more meaningful to forget about the details of the computer and instead gauge a problem's difficulty by how it "scales"—how much longer it takes to multiply, say, two six-digit numbers than two three-digit numbers, no matter how fast the machine. This relationship will hold true throughout technology's inevitable march. Driving 100 miles at a constant velocity will always take twice as long as driving 50, whether you are in a BMW or a VW Bug. The only way to break that barrier would be to find a wormhole to tunnel through, a rip in the fabric of space-time.

Simple tasks like driving steadily from point to point or counting the number of objects in a box are said to scale linearly. As the problem gets larger, the solution time increases at an even pace. If it takes 3 times longer to count 30 objects than 10 objects, then it will take 4 times longer to count 40 and 5 times longer to count 50. Draw the relationship on a graph and you get a straight slanting line. As the problem grows bigger, the solution time rises, but the increase is so gradual that a computer, and even a human, can handle larger and larger numbers with relative ease.

In the example here, a linear relationship is stated by the equation (or "function") $y = 2x$. Functions are like abstract little machines. Insert a number, x, into the input, press a button, and the product, y, emerges from the other end. According to the graph, a problem with a hardness rating of 2 (this might be the number of digits on which to perform some arithmetical operation or the number of items in a list to sort) takes 4 time units to solve (2×2). One with a hardness of 4 takes 8 time units, and one with a hardness of 10 takes 20 time units. Double the size of the problem and you double the solution time.

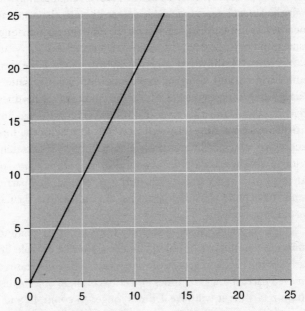

Figure 4.2. A graph of a simple linear relationship, $y = 2x$. As the input, x (indicated on the horizontal axis), increases, the output, y, rises gradually. When x is 10, y is 20.

Somewhat harder are problems that scale according to the power of some number. Suppose the relationship is $y = x^2$. Now the line on the graph rises more rapidly. A problem with a hardness of 2 still takes 4 time units, but one with a hardness of 4 takes 16 (4 squared). And so the acceleration continues: Problems with a hardness level of 10 take 10^2 or 100 time units, and those with hardness 20 take 20^2 or 400. Mathematicians call these "polynomial functions." For higher powers, like x^3 or x^{17}, the line rises even faster. (Download a graphing calculator from the Internet and you can play with the equations, developing an intuitive feel.) Even so, polynomial scaling is relatively tame. Unless the powers are enormous, the burden can be managed.

For the most interesting problems, the difficulty increases far more severely. It scales exponentially. Solution time grows not by some number to the 2nd, 3rd, or any fixed power, but to the xth power, where x is the size of the problem to be solved. Consider a problem that scales according to the innocent-looking relationship $y = 2^x$. A hardness of 2 (2^2) takes four time units, a hardness of 3 (2^3) takes 8, and a hardness of 4 (2^4) takes 16.

So far this is a fairly leisurely pace, but note that each time hardness is increased by a single step, the solution time doubles.

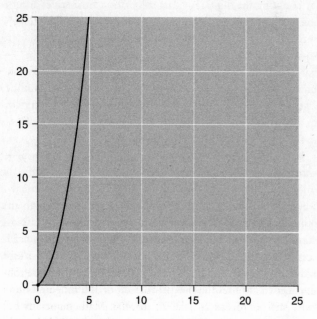

Figure 4.3. In this function, $y = x^2$, the output rises faster but is still well behaved. When x is 10, y is 100 (off the top of the graph).

Before long the graph begins to soar. Hardness level 8 takes 256 time units (2^8), and going up just one more notch, to level 9, takes 512 units. By now, the line is shooting practically straight up, running off the edge of the graph. To see hardness level 11— 2^{11} or 2,048 time units—the size of the page would have to be nearly tripled to two feet. And hardness level 12 (4,096 units) would take a four-foot-high graph . . . then, step by step, eight feet, sixteen feet, thirty-two feet. The graph of a problem of hardness 21 would stretch almost half a mile, and one of hardness 30 more than 200 miles. When problems scale exponentially, even modern computers are quickly overwhelmed.

For factoring a number into its primes, hardness is measured by counting the digits. Say that calculation time scales according to 3^x and that the time units are nanoseconds, billionths of a second. A 2-digit number would take 9 nanoseconds (3 squared) and a 10-digit number would take 59,049 (3^{10})— barely a flash—but a 20-digit number would take a full 3½ seconds, a 30-digit number nearly 60 hours, and a 50-digit number 23 million years. Try this with a calculator. It is astonishing how quickly these functions explode in size, with tiny little increases in the input causing huge eruptions.

In recent years mathematicians have found some clever shortcuts to make factoring faster. (With the best algorithm, the difficulty of the problem grows only "superpolynomially," which is still far worse than polynomial scaling though not quite as bad as a full-blown exponential explosion.) Using a complex method called the general number-field sieve, 292 computers in the Netherlands, Canada, the United Kingdom, France, Australia, and the United States took a little more than five months to find the two prime factors of a 155-digit number—still the record as of 2002. Because of the importance to cryptography, a security company has offered $200,000 to the first person who factors a 617-digit number. No one is holding

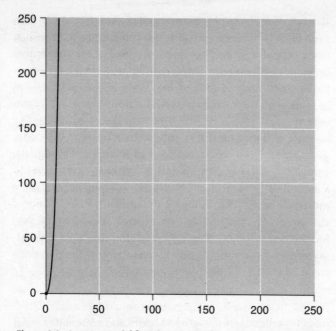

Figure 4.4. An exponential function, $y = 2^x$. The curve starts out slowly but soon is soaring almost straight up. If x is 10, y is 1,024. (The scale here has been compressed to emphasize the steepness of the curve.)

their breath. It is sometimes estimated that factoring a number 400 digits long would take a powerful supercomputer billions of years.

Forecasts like this are always risky. No one can predict how fast computers will eventually become. The point is that it doesn't much matter. Even if the estimate turns out to be off by a factor of a million, it would still take thousands of years. If even that turns out to be overly pessimistic, then just add a few digits. With exponential scaling the difficulty will soar out of sight again.

Of course, if you wanted a faster answer, you could enlist the services of millions of supercomputers working side by side. With billions of supercomputers you could factor the 400-digit number in a few years. But that is just replacing one exponential explosion with another. Shrinking the processing time would require the acreage of computers to burgeon beyond belief, and at some point there would be room for nothing else in the solar system.

There still may be that ingenious trick, a secret algorithm gone unnoticed all these centuries, that would slice through the complexity and factor large numbers with ease—a shortcut like a wormhole through space. But so far factoring appears to be inherently intractable, bumping up against the very limits of the power of computation—or at least computations performed on computers obeying the deterministic laws of classical physics. Until very recently no one seriously thought there could be any other kind.

Imagine a Turing machine—and it will serve as a stand-in for any kind of computer—programmed to factor a number, encoded as a long, long string of Xs and Os. The machine's head shuttles back and forth along the tape, slavishly reading, writing, and erasing marks according to its preprogrammed rules. One by one it is trying out the possibilities, dividing each smaller number into the larger number and rejecting those that leave a remainder. Intermediate results are stored in the cells cordoned off for use as memory, erased and replaced as needed.

When all the machinations are finally done, the input tape will have been converted into the output tape: a list of the factors. To save time, many Turing machines could work side by side, each simultaneously testing one of the possible factors. But there is always the same tradeoff: To handle longer and longer numbers, either the processing time or the number of Turing machines must grow exponentially.

Now imagine instead a quantum Turing machine. Each mark on the tape could be X, O, or X and O at the same time. As with its classical cousin, the machine would manipulate the tape, shuffling and reshuffling the bits. But since these are quantum bits, or qubits, every one of the numbers to test would be encoded on the same tape simultaneously, in quantum superposition. All could be tested at the same time, in a single run.

Figure 4.5. A quantum Turing machine. Each cell can be O and X at the same time.

For a somewhat more realistic picture, the tape of quantum Xs and Os can be envisioned as a row of atoms each surrounded by an electron spinning clockwise or counterclockwise, 0 or 1. The read/write head might consist of a laser gun. Nudging an atom with a pulse of the right frequency will cause it to resonate, to ring like a bell. Strike it for just the right duration and it will reverse its spin, flipping from 1 to 0 or vice versa. Zap an atom for half that time and there will be a 50-50 chance that the electron will end up spinning one way or the other. It will be, in other words, in a superposition of the two possibilities: Φ, 1 and 0 at the same time. By administering the correct sequence of pulses, the laser will convert the input string—the simultaneous encoding of every one of the numbers to be tested—into the output string: the answer to the factoring problem.

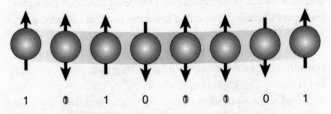

Figure 4.6. A somewhat more realistic picture of a quantum Turing machine. Each cell is an atom spinning up, down, or up and down.

The best way to get a feel for this strange idea is to hit it again and again from different angles. Each of the possible factors to test can be thought of as a quantum ripple, one piece of a complex wave packet. Computing the answer is equivalent to massaging the wave with the laser gun so that all the possible answers interfere with each other, the least likely canceling each other out and the most likely reinforcing each other. Finally the wave collapses to reveal the solution.

Or, changing metaphors again, think of the earlier example of a photon bouncing off the center of a mirror, so that the angle of its approach equals the angle of its departure. In the "many-paths" view, the particle can be thought of as trying out all the possible trajectories simultaneously. These possibilities interfere, reinforcing and undermining each other, finally yielding the likeliest path. Each trajectory represents one of the factors to test. The final one that emerges is the answer.

However you think about this quantum parallelism, the promise is machines that are immune to the exponential explosion. Remember that with a classical computer, each extra digit causes the factoring time to multiply. A number with 155 decimal digits took dozens of computers months to process; one with hundreds of digits could take billions of years—or billions of Turing machines.

With the quantum computer, handling longer numbers just means adding a few more qubits to the single row of atoms and then performing all the calculations in tandem. Solution time hardly increases at all. Here is another way to look at it: Each time a qubit is added to the string, it is the number of computations that can be done simultaneously that increases at an exponential rate. The explosion works in the machine's favor. When it comes to computation, quantum mechanics suggests a way for taking a shortcut through time.

5 | Shor's Algorithm

The strategy so far in these pages has been to edge steadily closer to an understanding of quantum computing with a series of increasingly better cartoons. The picture of the quantum Turing machine manipulating its tape of qubits comes close to capturing the essence. But look a little closer and some difficulties appear.

The classical Turing machine reads a cell, refers to its rule book, and acts accordingly: a 1 might be changed to a 0, or erased altogether, the read/write head then moving up or down the tape to the next designated locale. That works fine with classical bits. But a quantum Turing machine is processing quantum bits. And the most basic truth of quantum mechanics dictates that you cannot measure a qubit without destroying the superposition. The register that said 1 *and* 0 will randomly collapse into 1 *or* 0. The quantum juggling act will break down, the superposed bits falling to earth like so many dropped balls.

One way to get around the problem is to suppose that the read/write head itself is somehow part of the quantum system. Confronted with a cell that says Φ, the head will also enter a superposition, registering both 1 and 0. As long as the information doesn't leave the quantum world there has been no measurement—the delicate superposition won't be upset. But this gets harder and harder to visualize, since the head itself, being quantum-mechanical, can't be restricted to any one place at a time. It would have to be in an even more convoluted super-position, reading all the cells at once. Pressed too hard, the metaphor crumbles.

Drawing another step closer to what a real quantum computer would be like requires shifting perspectives again, looking at computing in a subtly different light. Take the image of the Turing machine, the cartoon version of classical computation, and dispense with the read/write head altogether, so all that is left is the tape.

The result is what mathematicians call a cellular automaton or CA—a deceptively simple device that can generate surprisingly intricate behavior. Start the tape with a pattern of cells colored black (1) or white (0). Then have each cell interact with its neighbors according to a strict menu of rules: "If the two cells to your left are black and the two to your right are white, then you must turn white also." Or "If the pattern to your left is white-black-white and the one to your right is black-black-white, then you must turn black." Any number of variations can be imposed. Once the device is set in motion, the cells interact with their neighbors and, at each tick of the clock, the pattern evolves kaleidoscopically, a scintillating pattern of black and white.

Watching a single row of cells blink rapidly on and off results in little more than dizziness. To make the patterns easier to fathom, practitioners of this arcane art program their computers so that each newly generated row is displayed beneath its progenitor. Pick the rules you want the system to follow, enter an initial sequence of cells, and the pattern unfolds, tick by tick, scrolling down the screen. (You can try this firsthand using the simulators available on several Web sites.) Some configurations quickly get stuck in a rut, settling into a dull repetitive cycle, stuttering ad infinitum.

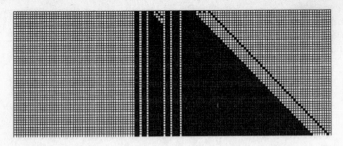

Figure 5.1. A boring cellular automaton. The pattern at the top fluctuates for a few steps and then settles into a dull routine.

Other automata churn out pleasingly symmetrical patterns.

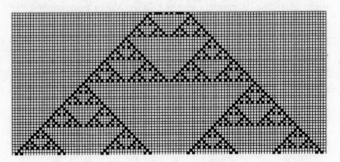

Figure 5.2. A symmetrical cellular automaton. Here the simple starting pattern generates a kaleidoscopic output that appears almost crystalline.

And some produce gibberish.

But a few seem to evolve endlessly according to some secret plan, generating intricate designs—the tension between order and chaos that we call "complexity."

Remember that computing can be thought of as taking a row of cells, the input, and converting it into an output: the pattern representing the answer to the problem. Thus, with the correctly crafted rules, a cellular automaton can be programmed to

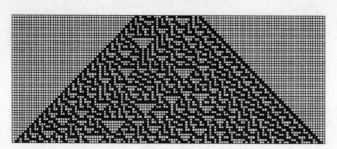

Figure 5.3. A disorderly cellular automaton. The pattern that emerges here seems to follow no rhyme or reason.

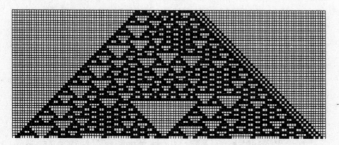

Figure 5.4. A very intricate cellular automaton. The pattern that is generated is orderly yet surprising—the hallmark of complexity.

compute. Suppose we want to make an adding machine. It would take as its input a row of cells colored black or white, 1 or 0, that might look like this: 0000000000110111, for 2 + 3. Then it would transform itself into a pattern interpreted to mean 5: 0000000000011111. (We're not using real binary here, just counting 1s.) A different, more complex rule set might conceivably take a pattern representing some number and convert it, after many generations, into its prime factors, in the form of a sequence of black and white cells. The device would act, in other words, like a Turing machine but without the need for a read/write head.

An engineer could think of many ways to realize so abstract a device. The cells might consist of a row of cards colored black on one side and white on the other. Backstage, the cards would be linked with an elaborate set of gears and pulleys that captured the proper dependencies: a card will be white if all three of its left-hand neighbors are black and the two outer neighbors on the right-hand side are white. But if the right most card is flipped over, then the middle one must change as well, along with the card on the far left. Or whatever. The example here is arbitrary. The possibilities are endless. In an electronic version, each cell could be a lightbulb wired to its neighbors, Geniac style, with a skein of logic circuits—AND, OR, and NOT gates. There is probably a way to do it with Tinkertoys.

Taking this design into the quantum realm requires what is by now a familiar leap: Decree that every cell can be black, white, or black and white at the same time. As with the quantum Turing machine, the cells might consist of atoms each with

Figure 5.5. A mechanical cellular automaton. Each card is black on one side, white on the other. Flip one to its opposite state and some of its neighbors must change as well.

an orbiting electron that can spin clockwise, counterclockwise, or both: 0 or 1 or Φ.

Suppose again that we want to break some number into its primes. First we zap each atom with a laser, putting the row of qubits into a superposition representing all the possible factors to test. Now that the data have been loaded, the procedure—dividing each smaller number into the larger number and looking for those that leave no remainder—is executed by applying another, more complex sequence of laser pulses, the equivalent of a computer program.

Let's call the atoms A, B, C, D, E, et cetera, and imagine that they are linked in a specific manner. If atom C is hit by a properly timed pulse it will change states, flipping from 1 to 0 or vice versa, but only if its two neighbors, B and D, are in the 1 state as well—the atomic version of a cellular automaton rule. Or D might flip only if its neighbors, C and E, are in opposite states. The algorithm might call for hitting atom E with a half-duration pulse, putting it into a superposition of 1 and 0. Choose the rules correctly and, in the end, the row of spinning atoms will be transformed into one representing the answers to the problem.

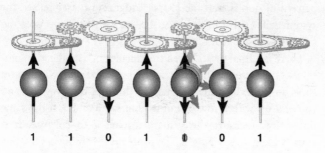

Figure 5.6. A quantum cellular automaton. Like the cards in the previous diagram, the atoms are interlinked (or "entangled"). Each one's state depends on that of its neighbors.

Unlike the head of the Turing machine, the laser in the quantum cellular automaton doesn't actually read the information on the "tape." It just fires off orders to be carried out. The qubits essentially read each other. The information stays inside the quantum realm, so there is no premature collapse.

For now don't worry about the details of atomic physics involved here. It is the underlying idea that is important. As illustrated in Einstein's thought experiment (the one he and Rosen and Podolsky tried to use to debunk quantum theory), subatomic particles can become entangled, their fates inextricably intertwined. If the electron on atom A is spinning up, then the one on atom B must be spinning down. Or they might be linked so that both must spin in the same direction. It depends on the kinds of atoms you choose. They don't even have to be adjacent. Atom A might be entangled with D, or E with B. In the classical cellular automaton, the cells are governed by an external table of rules. In the quantum CA, the rules are internal, consequences of the ways the qubits interact.

The result is a means of carrying out logical operations, the quantum equivalent of AND, OR, and NOT gates. There are no wires, like those in a classical computer, just the dependencies between the atoms. Given a problem to solve, coded as an initial pattern of qubits, and the proper program of laser pulses, the system will evolve to produce the answer, in happy isolation from the outside world.

There is a final twist. When the calculation is done, the solutions—the factors of the number being analyzed—will all be in superposition. Run the number 15 through the program and you end up with a row of qubits that say 3 and 5 at the same time. Now measure the system, forcing it to collapse. It is equally likely to say 3 or 5. The problem is complete. Dividing the factor, whichever one happens to fall out, into the larger number (here a regular pocket calculator or even a brain will do) yields the second factor. If there are more than two factors, the calculation can be run again. Each time the system ends in the same superposition and a measurement chooses one of the factors at random. Enough samplings will ferret them all out.

This image of a quantum CA is still just the inkling of an idea, an abstract plaything in which the details are smeared over in hopes of conveying an intuitive feel. In 1994, Peter Shor, a researcher at Bell Labs, took the airy notion of quantum factoring and made it more solid. He wasn't trying to actually construct a quantum computer from spinning atoms. He is a mathematician, not a physicist. But he showed that if one could be built—and nothing, remember, in the laws of physics seems to prevent it—then it could indeed be programmed to factor numbers. And, more important, he proved that such a machine would overcome the exponential explosion. As the numbers to be factored increased in length, the processing time would grow far more slowly than with any conceivable classical computer. Numbers that would otherwise be impossible to factor—it would take longer than the universe is expected to last—could be handled with ease.

Shor made his breakthrough using a touch of mathematical sleight of hand: Faced with a recalcitrant problem, mathematicians try "mapping it" onto an easier problem. If the two are equivalent, then solving one is tantamount to solving the other. According to legend, the German mathematician Carl Friedrich Gauss demonstrated this kind of prowess when an elementary school teacher saddled his class with some tedious busywork: adding up all the numbers between 1 and 100. As the other students sat laboriously summing $1 + 2$, then $3 + 4$, then $5 + 6$, and so on, Gauss saw that there was a better way: add $1 + 100$, then $2 + 99$, then $3 + 98$, and so on, until you reached $50 + 51$. But you wouldn't need to actually do all the additions. Each sum is obviously 101 and there are 50 of them, and $50 \times 101 = 5,050$.

He had invented a time-saving algorithm: add the first number to the last number (1 + 100), then multiply this sum by an amount equal to one-half of the larger number. Adding all the numbers from 1 to 1,000 the old way would take at least 10 times longer than for the numbers from 1 to 100. But with Gauss's algorithm you would simply multiply 1,001 by 500. To add all the numbers up to a million, just multiply 1,000,001 by 500,000. The procedure takes longer than for 1 to 10 but not 100,000 times longer. The algorithm scales much more slowly.

Shor, looking for his own shortcut, took advantage of a surprising fact well known to mathematicians. Because of some deep hidden connection in the web of mathematics, factoring can be mapped onto another, completely different problem: examining a sequence of numbers and figuring out its "period," how often it repeats. The sequence 123123123123, for example, has a period of three. It can be thought of as a wave:

Figure 5.7. A number wave. The repeating pattern has a period of 3.

To factor a number, you start by plugging it into a simple formula that generates one of these undulating streams. Encoded in its rhythm are the factors. The trick is teasing them out. Just as a wave of sound or a wave of light can be analyzed for hidden patterns, so can the number wave. Run it through the correct algorithm (a kind of mathematical filter) and the factors come dropping out. This will become clearer when the procedure is described in more detail below. For now just imagine a number refracted through some kind of mathematical prism, casting its factors on the wall.

When using this rather indirect method to solve a large factoring problem, an enormous number of calculations are still involved. The short explanation of what Shor did was to show that all these could be performed simultaneously, in quantum superposition. To put it another way, his algorithm created a quantum waveform representing every possible factor and then collapsed it to produce the answer.

That is the gist of the idea, to be carried into the next chapter (and it wouldn't be disastrous to just skip ahead). But with a little patience it is possible to peer inside the black box, appreciating the algorithm in more detail. The longer explanation, which makes the discovery all the richer and more concrete, involves playing with modular or "clock" arithmetic, a subject that may sound esoteric but is actually not much harder than telling time.

Normally the number system is envisioned as an endless row of integers: 0 1 2 3 4 5 6 7 8 9 10 11 12 13 14 15. . . . Adding 7 and 6 is a matter of starting at 7 and hopping 6 places to the right to land on 13. Modular arithmetic simply uses a circular number system like a clock face.

Figure 5.8. An ordinary clock. Adding together hours involves a different kind of arithmetic.

What is 7 o'clock plus 6 hours—13 o'clock? Wrong number system. To get the correct answer, start at 7 and move 6 places around the dial, landing on 1. In this system 7 + 6 = 1. And 10 + 4 = 2. Rather than moving your finger around the circle, you can get the answer with simple arithmetic: Add 10 and 4 the conventional way to get 14. Divide 14 by 12 (the number of numerals on the clock) and take the remainder, 2, as the answer. Mathematicians say that 10 + 4 = 2 (mod 12).

A mod 5 number system is a clock with 5 numerals:

Figure 5.9. A mod 5 clock has fewer numbers on its face.

Counting around the face, 4 + 5 = 4 (mod 5). Or add 4 and 5 to get 9 then divide by 5. It goes in once with a remainder of 4. Another way to say this is that 9, in our familiar number system, is 4 in the mod 5 system. What is 33? Start at 1 and count around the clock face 33 positions. You land on 3. So 33 = 3 (mod 5). For larger numbers, the counting becomes laborious. Just divide by 5 and take the remainder.

The first step to factoring with modular arithmetic is to create a special clock face for the problem, one with a number of digits equal to the number you want to break into primes. To factor 15, for example, you would use a clock with 15 digits on the dial.

Figure 5.10. The mod 15 clock has three extra numbers.

Then you use this mathematical contraption to generate a number wave. Analyzing its period, how frequently the undulations repeat, will reveal the factors. The procedure, though somewhat tedious, can be described as a sequence of very simple steps—just what one needs to make an algorithm.

First you arbitrarily pick a number that is smaller than the one you want to factor. (This sounds weird, but it works.) Let's choose 7. Then raise this number to the first power, the second power, the third power, and so on, translating each answer into clock arithmetic.

To begin: 7 to the first power is just 7. And, counting around the clock face from zero, the answer in mod 15 is also 7. (Or, as we've seen, we can divide 7 by 15 and take the remainder. Since 15 is too big to go into 7 even once, the whole amount remains.)

Now repeat the process, raising 7 to the second power: 49. Fifteen goes into 49 three times, leaving a remainder of 4. We now have the first two terms of the sequence: 7 and 4.

Continuing, raise 7 to the third power, $7 \times 7 \times 7$, which is 343. Fifteen goes into it 22 times with 13 left over. For $x = 4$ we get a remainder of 1 again, for $x = 5$ a remainder of 7.

Something interesting is happening. As we raise 7 to higher and higher powers, the sequence that emerges begins repeating

itself: 7, 4, 13, 1, 7, 4, 13, 1, 7, 4, 13, 1, 7, 4, 13, 1, . . . , ad infinitum. It is not the numbers themselves that are important here
but the rhythm—a wave with a period of 4.

Figure 5.11. Hunting for primes. In Shor's algorithm, a number
wave can be generated and then "refracted" like a light beam to
reveal the factors of a number.

Now the finale. To coax the wave into giving up the factors of
15, set it aside momentarily and do just a little more figuring.
First take the period of the wave we have just finished generating and divide it in half. Then raise 7, the arbitrary number
picked at the beginning of this arithmetical marathon, to that
power. So we halve 4 to get 2, then raise 7 to the second power:
$7^2 = 49$.

The last step (and this is really it) is to pick out the two integers on either side of 49—48 and 50. Now we compare each to
the original number 15, the one we are factoring, and see what
their largest common divisors are, the numbers that will go
evenly into both. For 48 and 15 the common divisor is 3. For 50
and 15 it is 5. And 3 and 5 are indeed the factors of 15. The
problem is solved.

We would have gotten the same answer through a different
route by choosing 2 or 11 for our arbitrary starting number.
Plugging either one into the algorithm would generate different
sequences of numbers, with different periods. But when the
analysis was done, the same factors, 3 and 5, would emerge.
(The procedure is not foolproof. Sometimes you get sequences
with an odd number of repeating elements—a period of, say, 5.
This causes problems when you get to the part where you're
supposed to cut the period in half. But a computer can keep trying different values until it gets one that works.)

Obviously it is a lot easier to factor a small number like 15 the traditional way, dividing it by 2, 3, and 4 to see which go in evenly. For much larger numbers, though, clock arithmetic can be a little faster than trial and error. Though the algorithms still scale exponentially, the slowdown is not quite so precipitous. The advantage is slight. With longer and longer numbers, the rapidly increasing load of calculations is still overbearing.

But that is for a classical machine. Shor suspected that with a quantum computer clockwork arithmetic would have an advantage: The many calculations needed to produce the number wave could be carried out simultaneously. Of course all the trial-and-error divisions of the old-fashioned method might also be done somehow in superposition (and it's perfectly OK to keep thinking of it that way). But there was a very good reason to follow Gauss's example and take the more indirect approach. Like waves of light or waves of sound, waves of numbers can be analyzed using a powerful technique called the Fourier transform (after the French mathematician Jean-Baptiste Joseph Fourier). Among the information that can be extracted is the period of the wave.

Shor knew that other researchers had recently stumbled across what seemed at the time to be little more than a curiosity: The Fourier procedure could be run on a hypothetical quantum computer. The technique is one best left inside its wrappings; unpacking it would involve too great a load. Suffice it to say that the Fourier transform essentially involves generating a bunch of test waves, each with a different period, and then trying them one by one to see if they match the wave you want to analyze. In a quantum computer all these probes could be applied simultaneously. Shor believed the technique could be used to rapidly extract factors.

Recall that the clock-face algorithm starts with an arbitrary number, smaller than the one you want to factor, and uses this as a seed, raising it to the first power, the second power, and so forth, converting the answers into modular arithmetic. This gives the sequence whose period will reveal the factors. Shor's algorithm starts by putting all these powers—the integers 1, 2, 3, 4, and so on—into quantum superposition. First, of course, they must be converted into binary: 0, 1, 10, 11, 100, 101, 110, 111, 1000, . . . And, as we've seen, all these strings can be represented simultaneously by a long row of spinning atoms. We will call this the "input register."

With all these numbers hovering together, the next step is to perform the calculations, raising the seed to these successively higher powers. Shor showed how this could be done with the quantum equivalent of AND, OR, and NOT gates. The atoms in the register are arranged so that some are dependent on others. If one is spinning up, another must be spinning down. Once the atoms are properly "wired" this way, the right sequence of laser pulses will carry out the clock calculation. And since all the input values are in superposition, the calculations needed to produce the output—the terms that form the number wave—can all be done at the same time.

Now comes the hardest part: the answers will appear—again all in superposition—in an "output register," also made of spinning atoms. (Without worrying about it too much, consider that each atom in the first register is quantum-mechanically linked, or "entangled," with a corresponding atom in the second register, in much the same way that the two photons in the EPR experiment are intertwined.) Suppose the quantum computer has just finished applying the series of pulses to factor 15. All the elements in the resulting number wave—1, 7, 4, 13, 1, 7, 4, 13, 1, 7, 4, 13—are now suspended in superposition in the output register.

All that is left is to figure out what the period is. Measuring the output register causes it to randomly collapse into one of the numbers in the sequence, say 7. That alone doesn't tell us anything very useful. But it so happens that since the two registers are entangled, the first one will now *partially* collapse—into

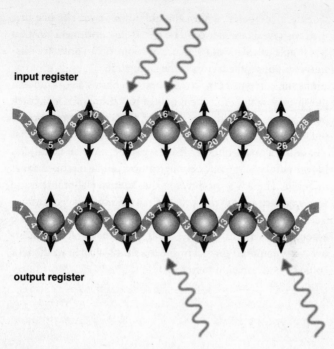

input register

output register

Figure 5.12. Shor's algorithm. In the first register, a row of atoms is put into a superposition representing all the integers, 1, 2, 3, and so on. With the proper sequence of laser pulses, this input is converted into the output, which appears in the second register.

a superposition consisting of all the input numbers that generated 7s when they were run through the clockwork mill: 1, 5, 9, 13, 17, 21, et cetera.

Notice that there is a pattern to these numbers: They are consistently spaced four units apart. They have, in other words, the very same rhythm as the output sequence. This is the number wave that will undergo the final measurement. The quantum computer fires off a final salvo of pulses that carries out the Fourier transform: Test waves of various frequencies are simultaneously compared against the unknown wave. The one that matches reveals the period. The rest is just arithmetic. With a few simple calculations (these can be done on an ordinary classical computer), the factors come falling out.

Having come this far, we can wrap up Shor's algorithm and put it away again. From here on out just think of it as a black box that uses quantum superposition to slice through the computational thickets, factoring any number in record time. What works for a small number like 15 can be done as readily on longer numbers, provided enough atoms can be recruited to act as qubits. There is a good reason that so many laboratories are working hard to pull this off: The payoff would be enormous. Recall that the fastest supercomputer, constrained to operate according to classical physics, could take eons to factor a number a few hundred digits long. With Shor's algorithm the task could be performed in minutes.

6 | Breaking the Code

If factoring were of interest only to pure mathematicians, Peter Shor's paper still would have caused a small sensation. It was fun for theorists to speculate about what one might do with a quantum computer. But until the factoring algorithm was discovered, there was no clear reason to actually build one, especially considering the heroic feats of engineering that would be required. Even if such a strange device could be constructed, it might not be able to do anything a Tinkertoy computer couldn't already do. Now scientists knew there was at least one problem, factoring, at which quantum computing would excel.

But there were more than abstract mathematical considerations involved. Many of society's secrets, from classified military documents to the credit card numbers sent over the Internet, are protected using codes based on the near-impossibility of factoring large numbers. Cracking one of the messages is equivalent to breaking a very long number into its prime components—something that was supposed to be impossible in any civilized length of time. An eavesdropper who managed to intercept a stream of protected information would be faced with a Herculean task: a series of trial-and-error calculations that could go on for millions or even billions of years. A working quantum computer would sweep that barrier away.

The system that suddenly became vulnerable is called "public key cryptography," the outgrowth of a centuries-old tradition of guarding secrets by methodically converting them into an unreadable form. All codes since ancient times are based on the same general idea. The information to be protected, called the "plaintext," is altered by some well-defined process to produce the encoded "ciphertext." Run the same process backward and the original message is restored. The coding scheme, or "key," might be as simple as replacing each letter with the next adjacent one in the alphabet—*a* becomes *b*, *b* becomes *c*, et cetera. "Quantum" would be disguised as "rvbouvn." Or each letter can be assigned to another at random. If $c = r$, $o = a$, $d = b$, and $e = x$, then "code" becomes "rabx."

Simple encryptions like this are fairly easy to undo, even without the key. If an eavesdropper intercepts a ciphertext and finds that its most frequent letter is x, then this is probably the substitution for e. The second most common letter in English is *t* and the third is *a*—more potential clues. Analyzing letter combinations yields additional information. Certain letters often appear as doublets—*ee, oo*—while others, like *w,* rarely do. A vowel can appear before or after most any letter, while consonants are more particular. The letter *q* is always followed by *u.* The letter *h* often precedes but rarely follows *e.* "Every letter in the English language has its own unique personality," as Simon Singh put it in *The Code Book.* The longer the encrypted message, the more likely it is to yield to analysis.

Another encoding technique uses transposition instead of substitution. The original letters of the message are retained, but they are rearranged according to a procedure known only to the sender and receiver. Maybe the first letter is swapped with the third letter and the fifth with the seventh, and so on down the line. Cracking the code is like solving a word-jumble problem, in which the jumbling has been done according to a precise plan.

While simple codes like these might be adequate for protecting notes passed in a school classroom, the secure transmission of state and financial secrets has inspired more devious schemes. The most impenetrable is to convert the message into a string of digits (with 01 for A, 02 for B, 03 for C, et cetera), producing a long number that can be disguised by adding it to a second number—one that has been chosen at random.

Transformed this way, and with the spaces deleted, the phrase "breaking the code" becomes

02180501110914072008050315040 5

Adding a randomly generated number produces the encrypted message:

plaintext	0218050111091407200805 03150405
	+
random number	3650972370743908278 12398753468

ciphertext	3869022481835315478 92901903873

The scrambled message can now be transmitted and converted back into the original only by someone who knows the right number to subtract. If this key is truly random, and if it is as long as the entire message, the code is unbreakable—so long as the key is used once and thrown away. If it is reused or a shorter key is applied repeatedly within the same plaintext, then an eavesdropper might be able to program a computer to scan for patterns, statistical regularities that can be exploited.

To make code breaking as hard as possible, more complex variations can be used. The first half of the message might be encoded using one key and the second half using another. Then the halves could be swapped and the whole number encrypted again. One of the main reasons computers were developed was to make this kind of many-layered ciphering faster and more efficient. A common encryption system used by governments and businesses around the world is called DES, for Data Encryption Standard. Messages are changed into numbers and then the numbers into binary code. The long sequences of 1s and 0s are chopped into sections, the sections are scrambled and combined, and the resulting number string is chopped, scrambled, and combined again—a process often compared to kneading dough.

There are many different ways to carry out the procedure. The particular recipe used to mangle a message is described by a number 56 binary digits long. This is the highly coveted and heavily guarded key. To encode a message the sender chooses one of the possible keys (there are 2^{56}, or more than 72 quadrillion) and enters it into the encryption program. The plaintext goes in one end and, when the kneading is complete, the ciphertext emerges from the other. When the recipient receives the message, he enters the same key into his copy of DES. The process is reversed and the original text is restored. For added protection, the algorithm can be run three times with three different keys, a procedure called triple DES.

No matter how deeply embedded the message, all the codes described so far have a fatal weakness: You must be able to communicate the secret key to the intended recipient without it falling into other hands. The reason people use codes in the first place is because they suspect that their transmissions could be intercepted. Why then should they assume the key is safe? Of course you could encrypt the key, but then you'd need a key for the key, ad infinitum. For centuries the hardest part about keeping secrets has been distributing and guarding the encoding and decoding instructions. Banks and government agencies dispatched human couriers with locked briefcases to fly around the world passing out the latest instructions. Since the keys had to be regularly changed to keep them from becoming compromised, the couriers' task was endless.

What is widely considered the biggest breakthrough in the

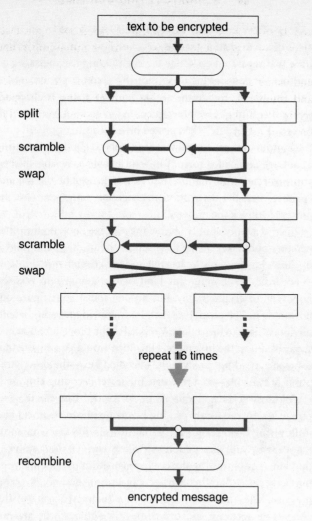

text to be encrypted

split

scramble

swap

scramble

swap

repeat 16 times

recombine

encrypted message

Figure 6.1. A simplified flowchart showing the steps in the convoluted Data Encryption Standard, which is used to protect sensitive information.

history of cryptography came in the 1970s, when mathematicians discovered an ingenious solution: how to make codes that have two separate keys—one for enciphering the message and another for deciphering it. To receive a secret communication, first you must provide the sender with your key. You needn't worry that it might be intercepted. In fact this "public" key can be broadcast far and wide, posted on a Web site or tagged onto the bottom of every e-mail you send. All it is good for is putting a message *into* coded form. You want people to be able to send you protected information. But as sole keeper of the second, "private" key, the decryption instructions, only you have the power to unlock the message.

To send a reply, simply reverse the process. Look up the other person's public key, use it to scramble the message and send it on, knowing that only he has the means to decipher it.

For years developing this kind of two-key system was the holy grail of cryptography. But no one could quite figure out how to do it. The procedure that unscrambled a message would obviously have to be somehow related to the procedure that had scrambled it in the first place. How then would you ensure that someone could not analyze the encoding key—the one that is publicly available—and re-create the secret decoding key?

The answer turned out to be factoring: You create your public key by randomly picking two large prime numbers and multiplying them together. (Actually this is done for you automatically by your encryption software.) The very long number that emerges is used by the coding algorithm to produce your public key, the one that anyone can use to send you a secret message. They just enter the key into their own copy of the encryption software and out comes the protected ciphertext. But unpacking the secret message requires knowing the two original primes, and those never leave the sender's computer. In theory, these two seed numbers could be re-created by breaking down the large number embedded in the public key. But that would require factoring it. With a long enough key, the task would be endless.

This powerful discovery, called the RSA algorithm, after its three inventors, Ronald Rivest, Adi Shamir, and Leonard Adleman, revolutionized cryptography. Finally it was possible to conveniently send secure messages without the risks that arise from distributing secret keys—ones that are used for both encoding and decoding. Only with Shor's discovery has a possible vulnerability come to light: RSA cryptography depends on the fact that computers are constrained by the laws of classical physics. If a quantum computer were put into service, security would be overturned. Once an eavesdropper had intercepted the encoded message, he would simply take the public key, readily available to anyone, and feed it to a quantum computer programmed to run the factoring algorithm.

Strong RSA cryptography commonly employs public keys that are 309 decimal digits long. Translated into binary, that means a string of 1,024 1s and 0s. So the interloper would need a quantum computer with 1,024 qubits—a row of approximately a thousand atoms or subatomic particles spinning up or down. (Other considerations, like circuits for error correction, would require additional qubits, but there is no need to worry about that now.) In the dark recesses of the code breaker's laboratory, the machine would try out all the possible factors simultaneously, in superposition, then collapse to reveal the answer.

The fact that a quantum computer could crack some of the strongest codes was reason enough to want one. The announcement of Shor's algorithm unleashed a flood of research grants from the Defense Department, the National Security Agency, the National Science Foundation, and other founts of federal largesse. But for theorists less interested in practical applications than in the scientific nature of this new kind of computing, Shor's algorithm seemed like just the beginning. They wanted to know what else could be done by replacing bits with qubits, harnessing the deep structure of the physical world.

Scientists already knew a few other things a quantum computer could do. It seemed almost axiomatic that one could be used to calculate the behavior of subatomic particles, Feynman's old dream. One quantum system would just be simulating another quantum system. (In a paper published in 1996, an MIT researcher named Seth Lloyd showed that the scheme was theoretically sound.) And it was clear that a quantum computer could generate a truly random number, a feat impossible for a classical machine. Computers are, after all, designed to be perfectly predictable. When they behave randomly they are discarded or repaired. If a stream of unpredictable numbers is needed—to run a dice-throwing program or to encrypt a message—programmers usually settle for what are awkwardly called "pseudorandom" numbers. Though produced by a completely repeatable and orderly process, they appear to be without pattern.

The process works like this: Some arbitrary input, derived perhaps from the current date and time or the user's most recent sequence of keystrokes, is used as a "seed." The numerical information is then fed into an algorithm that performs some calculation, spitting out a sequence of digits that have no easily discernible order. One of the first pseudorandom generators, suggested in the mid-twentieth century by the mathematician John von Neumann, is called the middle-square method. You start with an arbitrary number, square it, and take its middle digits. That is the first number in the sequence. You then square it and take the middle digits again. That is the second number. Suppose the seed is 156. The square is 24,336 and the middle digits are 433. Now square that to get 187,489 and take the middle digits again: 8,748. The next entries in the sequence are 5,275, 8,256, and 1,615.

The output certainly looks random. But the process that produced it is repeatable. Given the same input, the algorithm will always yield the same series of numbers. Suppose they are used to encrypt a message. By discovering the seed used to start the process, a code cracker could conceivably re-create an identical key. In fact, the middle-square method has been shown to be a poor generator of even pseudorandomness. More sophisticated methods rely on other techniques, like clock arithmetic. But at their root they all suffer from the identical problem: the same seed necessarily produces the same "random" numbers.

For most purposes, pseudorandomness is good enough. But occasionally scientists need a truly random number, and for this they can turn to quantum mechanics. Because of the indeterminacy of the subatomic world, a quantum computer could produce a number that is undeniably random. A row of 20 qubits will hold 2^{20}, or just over a million, binary numbers in superposition. Measure this wave function and it will collapse with equal probability into any one of these possibilities. Measure it again and again and it will spew forth an irreproducible sequence of unrelated numbers.

No one would want to build a full-blown quantum computer just to manufacture randomness. There are simpler ways. Recently a company in Geneva (its slogan: "True Randomness upon Request") began marketing a "plug and play" Quantum Random Number Generator. Inside the small box, which is about the size of a telephone answering machine, photons are shot at a semitransparent mirror, either bouncing back or passing through. The "choice," of course, is random, and the resulting bit stream can be fed into a PC through a common USB cable.

There are many possible variations. The ticks of a Geiger counter follow an unpredictable rhythm as they signal the random decays of radioactive nuclei. This process can be used to generate genuinely random numbers. In fact, with a bit of a stretch, a Geiger counter and its radioactive source might be thought of as a quantum computer dedicated to performing a single task: producing a patternless sequence of clicks.

It wasn't long before scientists found a way to expand quantum computing's narrow repertoire. In 1996 another Bell Labs researcher, Lov Grover, showed that quantum mechanics could be harnessed to speed up the kind of massive searches that strain the limits of computers mired in the classical world. Searching is the very essence of computing, whether it involves looking for a page on the World Wide Web, a name in a database, or the best chess move. To compete with a human oppo-

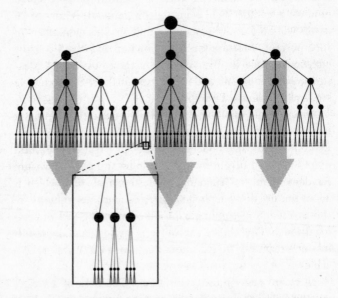

Figure 6.2. A portion of a simple game tree. For each move there are numerous possible counterresponses, and responses to the responses. . . .

nent, a computer must scan the board and analyze a vast number of possibilities. If a pawn is moved here, how would the opponent respond, and what would be the best response to the response—and the response to the response to the response? And what if the pawn is moved this way instead? The rapid proliferation of options can be laid out in a labyrinthine diagram called a game tree.

In the time allowed between moves, the computer explores the dense thicket as broadly and deeply as possible and then compares the various outcomes, picking the one that seems to leave it with the greatest advantage. The faster computers become, the more levels of the maze they can traverse. But they still must leave all but a fraction of the territory unexplored. Claude Shannon, the mathematician who invented information theory, estimated in 1950 that there are some 10^{120} possible variations in a typical chess game—far more than there are atoms in the universe. The technological advances since Shannon made his calculation are of little solace. Even supposing a computer could consider a trillion of the possibilities every second, it would still take more than 10^{100} years to completely analyze a game. The universe has existed only an insignificant fraction of that time.

Here again is a case where even the fastest classical computers are hampered by their serial nature, the need to deal with data one chunk at a time. The most powerful chess-playing machines, like IBM's Deep Blue, exploit a modest amount of parallelism, carrying out multiple operations simultaneously. When the computer beat chess champion Garry Kasparov in 1997, it drew on 256 processors working in tandem to examine 60 billion possibilities during the three minutes between moves. Think of 256 mice ganging up to explore a maze. That's better than one, but there was still a staggering amount of serial, step-by-step number crunching going on, and vast regions of the search space that had to be ignored.

What if a computer could explore the labyrinth by sending legions of mice running down every single branch, searching them simultaneously in quantum superposition? This is the possibility that has been opened up by Grover's algorithm. In the Kasparov game, each of Deep Blue's serial processors considered about 25 million positions before making each move, weighing and discarding them one by one. Because of the power accorded by quantum mechanics, the number of positions a quantum processor of equal speed could conceivably search in that time would be 25 million squared, or 6.25×10^{14}. Put 256 of the devices together into a quantum Deep Blue and it would search more than 10^{17} positions in three minutes—10 million times more than its classical counterpart.

The uses for a high-speed quantum search engine would be limitless. Suppose you have an alphabetical list of everyone in the world, with each name matched with a phone number. Looking up a number by name is easy, but if you have only a number and want to know who it belongs to then you, or your computer, must comb through the list item by item. If you're lucky you will chance upon the correct entry right away. If not you might have to search until the bitter end.

On average, you will have to scan through half of any list to find an item: it is equally likely to be in the top or the bottom. With two processors you can start at each end and cut the time in half. With four processors you can work four times as fast. But with a quantum computer, loaded with Grover's algorithm, you could examine each number simultaneously, in superposition.

That is the gist of his approach, enough to confer a sense of how this new technology might vastly speed the ability to sift through piles of unsorted information. Using a long string of qubits, every single item to be examined could be held hovering in superposition. Massage the configuration with the correct sequence of laser pulses—Grover's algorithm—and all the data would be analyzed in a single swoop. But like Shor's factoring technique, Grover's discovery invites a closer look. A few layers of wrapping can be stripped from the black box. The details are fascinating and again require nothing beyond simple arithmetic and a tolerance for following the steps of the algorithmic recipe. Along the way comes a deeper feeling for the power of quantum mechanics.

So far in this book, taking a bit of license, we've been describing the idea of a quantum superposition metaphorically—as a complex of wavelets each representing one of the many possible outcomes of some event. But it's a little more complicated than that. For an electron orbiting around a nucleus, each wavelet can stand for one of the many positions the particle might assume if it was measured. Each of these possibilities is described by a number that gives the wavelet's height, or "amplitude." Think of it as the loudness of the wave. It might seem natural to suppose that this figure describes the probability—one in a hundred, for example—that the electron will appear in a certain place. But actually the amplitude gives a different quantity: the square root of the probability. To get the probability, you must multiply the amplitude by itself. So if the height of the wavelet is $1/10$, the probability of the event it is describing is found by taking the square: $1/10$ times $1/10$ is $1/100$—one chance in a hundred, or 1 percent.

This connection between amplitudes and probabilities is more interesting than it might sound. While probabilities are always positive numbers (it would be meaningless for something to have less than a zero chance of happening), amplitudes can be positive or negative. A wave extending peaklike above the horizon has a positive amplitude. But it can also dip below the line, forming a valley or trough, a negative amplitude. If, on a scale of 1 to 10, the probability of an event is 9, the amplitude of the wavelet describing it might be 3 or -3. Either number squared gives the same result.

The crucial implication—and it is from this that a great deal

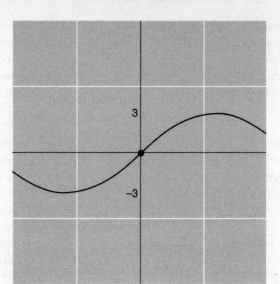

Figure 6.3. Laying odds on a quantum event. An amplitude of either 3 or –3 (on a scale of 1 to 10) means a probability of 9.

of the weirdness of quantum mechanics flows—is that positive and negative amplitudes can come together, peaks overlapping with valleys, and cancel each other out. This is precisely what is happening when a superposition—this complex of many possibilities—collapses to reveal a single outcome. When a particle, a packet of shimmering potentiality, strikes a barrier or is otherwise measured or disturbed, negative wavelets combine with positive wavelets, self-destructing and leaving the wavelet describing the lone possibility that is realized in the observable world.

Grover's algorithm relies on this phenomenon to rapidly search a repository of information. First all the entries are coded as 1s and 0s and placed together in quantum superposition. Without worrying about the details, recall that this can be done by using laser pulses to flip the spins of a row of atoms this way and that. The result is a tangle of quantum wavelets, one representing each item. Then the system is tweaked so that wavelets with positive amplitudes cancel out those with negative ones. Finally all that is left is the wavelet that represents the sought-after item in the database.

Those are the bare bones of the idea, to be fleshed out with a simple example. Begin with a short list of 16 bicyclists and their rankings (rendered in binary, of course) in a road race. The winner, Gina, is 00001. Sixth place (number 00110) goes to Paul; last place (number 16, or 10000) to Lolly.

Amy	00010	Ian	00101
Betty	00111	John	01000
Carrie	00100	Katherine	01110
David	00011	Lolly	10000
Ely	01101	Marianne	01100
Frank	01010	Nina	01001
Gina	00001	Oliver	01011
Harry	01111	Paul	00110

It is easy to glance at the list and see that Marianne ranked twelfth (01100) and Nina ninth (01001). The names are ordered alphabetically. But if you want to know which cyclist came in, say, fourteenth, you have to work a little harder—not much harder since this is a short list, but imagine if there were a hundred or a thousand names to pick through.

With Grover's algorithm, quantum mechanics can be used to speed the search. With just a few qubits, each in a superposition of 1 and 0, we can represent all the rankings, from 00001 to 10000, simultaneously. With a few more qubits, each number can be attached to a name. Here just the initial will do: A00010 (Amy), B00111 Betty ... (the letters, of course, would be numerically encoded). To put it another way, the list of cyclists and their rankings can be described by a packet woven from 16 wavelets each of equal amplitude. At this point, measuring the complex waveform would cause it to collapse and yield any one of the 16 possibilities. Thus the probability of each term appearing is 1 in 16, and the amplitude of each wavelet is the square root of $\frac{1}{16}$, which is $\frac{1}{4}$ or $-\frac{1}{4}$.

What has been described so far is just another random number generator. There is no control over what comes popping out. But suppose we are searching the list for a particular item, say the cyclist who finished ninth, or number 01001. Before any measurement had occurred, Grover's procedure would simultaneously examine all 16 entries (it could just subtract 01001 from the portion representing the ranking and see which one gave an answer of 00000). When it found the right match, it would amplify its wavelet, and juxtapose the others, peak to trough, so that they canceled each other out. The new waveform that emerged from this quantum massaging could now be measured. And it would collapse with very high probability to yield the item we are looking for: N01001. Ninth place went to Nina.

That still may seem a little vague. Exactly how are these complementary wavelets mixed and matched to produce the correct answer? Without some deeper mathematics, a certain amount of handwaving is inevitable. But, keeping with arithmetic, it is possible to peel off one more layer. To keep details to a minimum, consider a database of only 4 entries, 00, 01, 10, and 11. Place them in superposition (all it takes is two atoms spinning up and down). Each now has a one-in-four chance of being chosen, so the amplitude of each wavelet is the square root of $\frac{1}{4}$: $\frac{1}{2}$ or $-\frac{1}{2}$. At first all the amplitudes (one for each item) are positive.

Suppose we are searching for the entry 10. By dispatching the right sequence of pulses (and here we'll take it on faith that such a sequence exists), the computer manipulates the wavelets, simultaneously examining them for a proper match. (Again it can just subtract 10 from each entry and see which leaves an answer of 00.) Once it is found, the computer unleashes a sec-

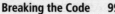

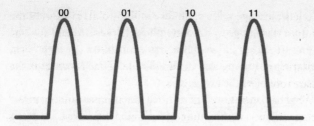

Figure 6.4. A quantum search. Each of the four items in the list is represented by the crest of a wave.

ond sequence of pulses that inverts that wavelet's phase, converting the peak into a trough, changing the amplitude to $-\frac{1}{2}$:

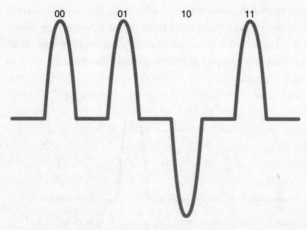

Figure 6.5. In the next step of the search, one of the waves is inverted.

This manipulation will not cause the superposition to collapse. Whether the amplitude is $\frac{1}{2}$ or $-\frac{1}{2}$, the probability is still $\frac{1}{4}$. So, from the point of view of an outside observer, nothing has changed. No information has leaked outside the quantum sphere, so the delicate superposition remains intact.

Next the computer issues a sequence of pulses that results in taking the average value of all the amplitudes. This is no different from averaging any list of numbers. First add them together: $\frac{1}{2} + \frac{1}{2} + (-\frac{1}{2}) + \frac{1}{2}$. Two of the terms cancel out (their wavelets overlapping), leaving $\frac{1}{2} + \frac{1}{2}$, or 1. Then divide 1 by the number of items, 4, to get the average: $\frac{1}{4}$.

Finally an operation is performed called "inversion about the average." This is a little more complicated but is still just plain arithmetic. First the algorithm is used to calculate how far above or below the average each wavelet is. Then it flips it by the same amount in the other direction. This is easiest to grasp by just doing it: The waves with $\frac{1}{2}$ amplitudes are $\frac{1}{4}$ *greater* than the average ($\frac{1}{4} + \frac{1}{4} = \frac{1}{2}$), so you take them *below* the average by the same amount: $\frac{1}{4} - \frac{1}{4}$ is 0. These waves then are flattened out. Now note that the wave with $-\frac{1}{2}$ amplitude is $\frac{3}{4}$ *below* the average, so take it *above* the average by the same amount: $\frac{1}{4} + \frac{3}{4} = 1$. The wave flips to the top of its axis and doubles in height. After the operation is completed the amplitudes are 0, 0, 1, 0.

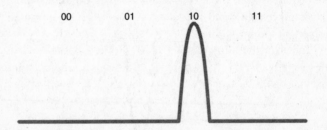

Figure 6.6. After a bit of massaging, we're left with a single crest representing the item we are looking for.

Now we just have to get the information out of the system. Recall that amplitude—height—is just the square root of the probability that the corresponding event will occur. So the probability that the measurement will collapse the wave into any one of the three incorrect terms is found by squaring their amplitudes: $0 \times 0 = 0$. The probability of choosing the correct term is $1 \times 1 = 1$. Odds of 1 out of 1 are as good as they come. The superposition will converge on the right answer. Searching larger databases would require repeating the steps of the algorithm again and again. With each round, the wavelet representing the item to be retrieved gets bigger while the others shrink toward zero. That is the important point.

Like Shor's factoring algorithm, Grover's method for searching seems rather convoluted. But he showed that there is a payoff. Instead of having to look through half the items of a database, as a classical computer generally must do, a quantum device need only comb through the square root of the number of items. For 16, that means considering 4 items instead of 8 to find the answer. That seems a pale advantage until you consider that for 1,000,000 items, a classical computer would, on average, have to look at 500,000 of them as opposed to only 1,000 with Grover's quantum algorithm. How much hardware would be required? Two to the 20th power is about a million, so the basic processing could be done with a row of 20 qubits, an object far too small to see.

The algorithm doesn't provide an exponential speedup like Shor's, but it still confers a huge advantage. Imagine a chess-playing machine that would beat not just Kasparov but any conceivable successor to Deep Blue.

The algorithm could also be used for code breaking. Suppose an interloper somehow obtained a scrap of ciphertext that had been created with DES, and that he also managed to intercept the corresponding portion of the original unencoded message:

Meet me at seven on 4th and Broadway
    wfhqyt1lk546nas32850lnmak2300ohn

Could he use this to re-create the key? The first step would be to enter the plaintext into his copy of the DES software. But then he'd have to try out all 72 quadrillion (thousand trillion)

DES keys to find the one that produced an identical section of gobbledygook. This could be used to unscramble future communications.

In 1998 a privacy group called the Electronic Frontier Foundation showed this was possible using a classical computer that, given the first 24 characters of plaintext, searched 88 billion keys a second, deciphering a short DES-encrypted message in only 56 hours. Half a year later they cracked a message in a little more than 22 hours by recruiting help from a worldwide network of 100,000 PCs, working in parallel over the Internet. Impressive as these feats were, Grover's algorithm, which would have to search only the square root of the total number of keys, might unravel a DES transmission far more rapidly.

What you would need is a quantum computer with two entangled registers—two rows of spinning atoms that are linked quantum-mechanically. First put all the possible keys in superposition in the first register using 56 qubits (remember that 2^{56} is 72 quadrillion). Then run a laser-pulse version of the DES algorithm, using each of the keys to convert the scrap of plaintext into its corresponding ciphertext. Since this is a quantum computer, all these calculations would be done simultaneously. The result would be all 72 quadrillion possible ciphertexts hovering in superposition in the second register. And since the two registers are entangled, each of these encodings is linked with the key that produced it.

If the system was measured at this point, it would randomly and unhelpfully collapse to produce any one of the 72 quadrillion pairs. But first use Grover's algorithm. The quantum computer examines the entries simultaneously and picks out the ciphertext we are seeking, which is linked to the key that produced it. Step by step, this wavelet's amplitude is inflated while the others are squeezed down. Finally a measurement will produce the correct answer.

.   .   .

In the absence of a working quantum computer, inventing algorithms like this is an act of pure abstraction, as well as faith. As Grover once put it, he and his colleagues are "writing the software for a device that does not yet exist." But slowly that is changing. In the last few years, physicists have finally coaxed a small number of qubits to carry out the first quantum calculations. Some experiments have actually run scaled-down versions of Grover's and Shor's algorithms—too simplified to be useful but serving as evidence that the basic ideas are sound.

These toy demonstrations are a small step toward practical machines that would break codes or play chess. But Grover warns against erring on the side of pessimism. In March 1949, he notes, an article in *Popular Mechanics*, describing a state-of-the-art computer called the Eniac, speculated on what lay beyond: "Where a calculator like the Eniac today is equipped with 18,000 vacuum tubes and weighs 30 tons," the writer predicted, "computers in the future may have only 1,000 vacuum

Figure 6.7. The Eniac (U.S. Army photo)

tubes and perhaps weigh only half a ton." To commemorate the fiftieth anniversary of the half-million-dollar machine, the supercomputer of its day, a group of electrical engineering students at the University of Pennsylvania duplicated its circuitry on a silicon chip measuring 7.44 by 5.29 millimeters.

7 | Invisible Machines

Just as the words in a language are intertwined in a great self-referential web of meaning, so can AND, OR, and NOT gates—the vocabulary of computation—be defined in terms of one another. If you are clever enough, you can make an AND gate with an OR and three NOT gates. In fact, with something called a NAND gate (an AND followed by a NOT) you can build any gate at all. (An AND gate will say 1 if both its inputs, A and B, are 1; otherwise it will say 0. With a NAND gate the output is then inverted. Feed in 1 and 1 and out comes 0.) Details aside, the important point is this: If all you had in your conceptual toolbox were NAND gates, you could still build a computer as powerful as a Turing machine. The NAND gate is what computer scientists call a "universal constructor."

A good starting point for designing a quantum computer, laying out a schematic for a working machine, is to find one of these all-purpose devices—a gate from which all others can be made. If you can produce just this one building block, it can be multiplied and combined to make any possible circuit.

As it turns out, quantum gates are trickier to put together than classical ones, and not just because they have to be made from single atoms or subatomic particles. There is an additional requirement: Quantum computation, unlike classical computation, must be fully reversible. To see what this means, consider what happens inside a simple pocket calculator. Punch in 2 + 2 and the output is 4. But along the way, the path to the answer has been destroyed. If you come upon a calculator displaying 4, you have no way of knowing whether the input was 2 + 2 or 1 + 3 or (237 × 558)/2 − 66,119. The registers that temporarily held that information have been cleared. The bits have disappeared from the system, shed into the surroundings as infinitesimal puffs of heat. The calculation is *irreversible;* you can't go back again.

The same kind of electronic amnesia occurs at the level of each individual gate. While a NOT gate is clearly reversible—if it says 1, you know that its input had to be 0—the OR gates and the AND gates operate only in a single direction. OR will say 1 if its input was 10 or 01 or 11. Given its output it is impossible to know what originally went in. The history has been erased.

By now it shouldn't be surprising that what holds true in the macroscopic world of classical physics is contradicted in the quantum realm. If a quantum calculator said "24," you would be able to reverse its computational gears and find out what the original calculation had been. The laws of physics demand that this be true: Any interaction between subatomic particles must be symmetrical in time. Run the reaction backward and the original state will be restored. If two particles interact to produce a third particle, then it must be possible for the third particle to decompose back into the original two.

A grander way of saying this is that in quantum systems information is always conserved, histories are never erased. If logic gates are made from individual atoms or subatomic particles, as they would be in a quantum computer, then the calculations they perform must be reversible as well. Manipulate the particles as much as you like to solve a problem. But once you have the answer, theory requires that there be a way to work back to the original question. It must be possible to go from

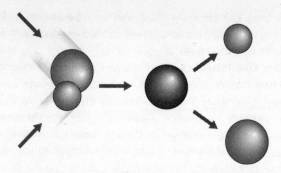

Figure 7.1. A reversible reaction. If a pair of particles can collide to produce a third kind of particle, then it must be able to break back down into the original two.

input to output and then back to input again. Any design for a quantum computer must take this truth into account.

When you think about it, this makes a lot of sense: Information leaves a quantum system only when it is measured or observed—or, more generally, when it interacts with the outside world. The superposition collapses, the calculation comes undone. As it works through a problem, a classical computer can discard the intermediate results—the computational leftovers—and keep chugging right along. In a quantum computer every scrap of information must be kept carefully isolated inside until the computation is complete. Otherwise it will fail.

There are good reasons that designers of classical computers don't worry about preserving a detailed history of the billions of rapid-fire calculations occurring every second. It would take skeins of extra circuitry to store what, in this case, amounts to useless information, computational trash. But there is a price for this profligacy. In the early 1960s, a physicist named Rolf Landauer proved that every time a bit is erased from a register, a minimum amount of heat is dissipated: wasted energy. And heat must be carried away from the processing chips to keep them from malfunctioning. Hence we are cursed to listen to the droning fans inside our computers, and behemoths like Q must be outfitted with industrial-grade cooling stacks.

Right now computers, like all machines, are so inefficient that the energy lost from erasing bits has hardly been worth worrying about. Far more heat is produced by the friction from electrons squeezing through the tiny wires or from the motors turning the hard disk drives. As chips get smaller and the wiring more densely packed, the heat becomes that much more intense. Engineers try to find ways to quickly disperse it or to lower the amount that is generated in the first place. But even if they could eliminate the motors and make chips with superconducting wires that offer no resistance whatsoever to electrical flow, they would still be confronted with Landauer's Principle: Every time a bit is erased an irreducible amount of heat is produced. The laws of physics require it. The only way to eliminate the heat is to avoid erasing the information in the first place, and that means saving all the intermediary results. As circuitry continues to shrink and cooling technologies are exploited to their limits, engineers are beginning to confront the problem of how to eliminate this final source of waste.

Thus, long before the current wave of interest in quantum computation, a farsighted few began thinking about how to make reversible gates for a future breed of classical computers—gates that do not throw away information. Given the output, you can know what the input must have been. A whole subfield called "reversible computing" has emerged in which circuits are designed, and sometimes constructed, that preserve every step of the computation. There are things called Fredkin gates and Toffoli gates, named after their inventors. There are imaginary designs for "billiard ball computers," in which computations are carried out by perfectly elastic balls ricocheting across a frictionless plane, a game that can be played forward or backward. The work is not all theoretical. A few energy-efficient experimental chips, made entirely from reversible components, have been assembled. And some circuits have been incorporated into laptop computers to save battery power.

These are all classical, macroscopic devices, with each gate made from billions of atoms. But the same ideas are now being used to design single-atom switches for quantum computation—gates that must, by their very nature, operate in both directions. Most important of all, physicists have found reversible gates that can serve as universal constructors. With just two kinds of operations, a single-qubit rotation (the ability to flip an atom between 1 and 0, or put it somewhere in between) and something called a "controlled NOT" gate, one can design any quantum computer at all.

We've seen time and again how to carry out a single-qubit rotation: A laser pulse of just the right frequency and duration flips an atom between 1 and 0. (Apply the pulse for half that long and the atom will be placed in superposition: Φ.) A controlled NOT gate is a little more complex, involving two atoms. The first is a control switch. If it is in the 1 position, then the second atom acts like an ordinary NOT gate, inverting whatever input it receives. Hit this atom with a pulse meaning 1 and it will flip over to 0—but only if the first atom is in the 1 position. If instead it is registering 0, then the neighboring NOT gate is turned off. Give the second atom either a 1 or a 0 and the information passes through unchanged.

It's easy to show that the action is reversible: Run the gate backward and the input is restored. Suppose that the output is 01. (See the top example in figure 7.3.) The first bit is the control signal. Since its only purpose is to either activate or deactivate the NOT function, it always passes through the gate unchanged. And since it is 0, we know that the NOT function was turned off. The second bit, in this case 1, also must have passed through unaltered. Thus we can deduce that the input, like the output, had to be 01.

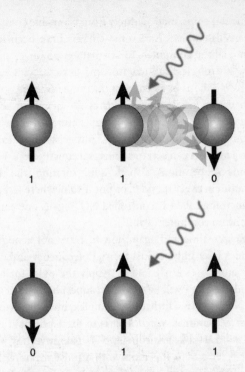

Figure 7.2. A controlled NOT gate. If the first atom is
in the 1 position, then a well-timed laser pulse
striking the second atom will reverse its value. But if
the first atom is in the 0 position, the pulse will be
ignored.

Now suppose instead (as in the bottom example of figure 7.3)
that the output is 11. Here the control signal is 1 so we know
that this time the NOT function was activated, causing the second bit, also 1, to be reversed. Hence the input had to be 10. In
either case, we can take the output of one controlled NOT gate
and feed it through a second one, restoring the original input.

So there we have it: a gate that is fully reversible, in accor-
dance with quantum mechanics, and universal—it can act as a

NOT function turned off

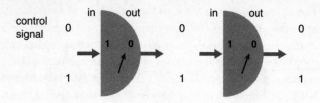

NOT function turned on

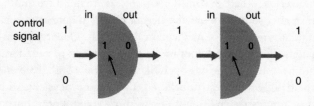

Figure 7.3. Reversible logic. The upper bit entering the contraption sets the switch to either 0 (off) or 1 (on). This affects how the bottom bit is processed. In the first example, the data pass through unchanged. In the second, the NOT function is activated. Either way the computation is reversible. Feed the output back into the gate and the input is restored.

building block to make any quantum computer. On paper, that is. The next step in the development of this new technology was to show that a quantum controlled NOT gate could be realized in the physical world.

The collision of ideas leading to the first quantum switch came in 1994 at an international meeting on atomic physics organized by the Boulder, Colorado, laboratory of the National Institute of Standards and Technology. One of the missions of NIST is to tell the nation what time it is. The most accurate means yet discovered involves using cesium atoms as pendulums, their tiny pulses acting as the ticks of an atomic clock. These timepieces are accurate to within a second over 15 million years. But that is still not considered good enough. For a civilization that depends on precise timing to track spacecraft, coordinate telecommunications signals, and calibrate physics experiments, it is impossible for a clock to be too exact. And so NIST's scientists try to keep abreast of the latest developments in atomic physics and quantum mechanics. Among the speakers invited to the meeting was Artur Ekert, a theorist at Oxford who is one of the leading promoters of quantum computation.

Ekert was there to spread his gospel. He found it frustrating that, after so many years, the elegant notion of quantum computation remained a mathematical curiosity. Quantum computers did amazing things in the platonic realm of pure idea, but he wanted to see one operating in the real world. If anyone could pull off this feat, he believed, it was atomic physicists with their impressive ability not just to think about atoms, but to manipulate them as well. In a widely publicized demonstration of this skill, scientists have used a device called a scanning tunneling microscope to arrange 35 xenon atoms to spell out "IBM"—an early example of "atomilist" art.

After going over some basic ideas—qubits, quantum parallelism, and so forth—Ekert laid down a challenge. All that was really needed to launch the revolution, he told the audience, were quantum logic gates, atomic-sized devices that, given a pair of bits as input—00, 01, 10, or 11—would perform some kind of computation. In fact, all that was needed to start with was a single controlled NOT gate, from which all other gates could be built. A working quantum computer would require many such devices all operating in harmony. But the task had to begin somewhere. Isolating an atom or two for the fraction of a moment that was required to shuffle a couple of bits would be a proof of principle, a sign that the promises of quantum computing were at least within the realm of the possible.

In the audience were two physicists from the University of Innsbruck in Austria, Juan Ignacio Cirac and Peter Zoller. After returning to Europe they began laying out the blueprint for a device that would meet the demands of Ekert and his colleagues. The technology needed to carry out this feat, they realized, already existed, in the form of an elaborate piece of laboratory equipment called an ion trap.

For years experimenters working in different corners of physics had developed techniques for capturing and manipulating single ions—atoms with a small electrical charge. In its normal state, an atom is electrically neutral, with the positively charged protons in the core exactly counterbalanced by a halo of negatively charged electrons. But if one of the electrons is stripped away, the balance is set askew. The atom gains a positive charge. Now it can feel the sway of electrical and magnetic fields, allowing experimenters to grab it and move it around.

By applying these forces just so, scientists can suspend a single ion inside a vacuum chamber. Then, to dampen the ion's motion, they bombard it from every direction with laser pulses, nudging it this way and that way until it is almost standing still—frozen within the clutches of the trap. (Since heat is simply random atomic motion, this technique is sometimes called optical cooling.) The outcome of these experiments is almost beyond belief: An individual atom—for so long just a symbol, a numbered square on Mendeleev's Periodic Table of the Elements—can be pinned like a butterfly on an examining table.

Cirac and Zoller saw that one of these trapped atoms would make a fine quantum gate. Suppose the ion had a single electron in its outer shell. If the electron was in the lowest possible rung of its orbit, its lowest energy state, this would indicate a 0; a higher energy state would mean 1. Hit the atom with a laser beam tuned to the right frequency and it could be flipped back and forth like a toggle switch (or suspended in between, Φ).

The next step would be to extend the idea to a row consisting of several ions, held side by side in the clutches of an electro-

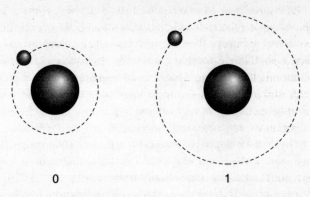

Figure 7.4. An ion switch. An electron in a low orbit is decreed to mean 0, a high orbit to mean 1.

magnetic field. The ions' similar charges would cause them to repel each other, while the waves of electromagnetism pushed them back together. Think of them as balls hanging from strings, little pendulums touching side to side. If one atom moves, the motion is picked up by its neighbors. They all rock back and forth in unison.

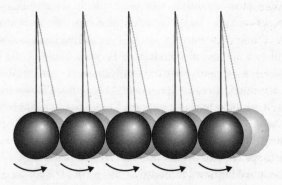

Figure 7.5. Rocking ions. Like pendulums they can swing in unison.

Of course atoms are not really balls. Since this is the sub-microscopic realm, the rocking motion is quantized. In a classical system, objects can vibrate across a smooth range of frequencies; a pendulum can swing at any speed. In a quantum system, vibrations, like everything else, can take on only discrete values, with nothing in between. (And something that is quantized must be carried by quanta; thus the "particles" of mechanical vibration, a category that includes sound, are called "phonons.")

The implications begin to become clear. Anything that can be in one of two distinguishable, black-or-white states can be used as a bit. The ground state, where the atoms are at rest, can arbitrarily be called 0, with the next higher state of rocking defined as 1. And the row of atoms can also be in superposition, rocking and not rocking at the same time. (Don't even try to picture that.)

The result of all this would be a quantum register with two separate ways of storing information—according to whether an individual atom (or, more specifically, its orbiting electron) is in its high- or low-energy state and whether the whole row of atoms is vibrating or standing still. For the final step of their argument, Cirac and Zoller showed that their hypothetical device could carry out logical operations, that it could compute.

Suppose, for example, that one of the ions is in its excited state, 1, and that it is rocking in the low-energy mode, 0. The result is a tiny register storing the information 10. Cirac and Zoller showed that with a laser pulse tuned to the right frequency, the electron could be smacked back down to its ground state, 0, while, *at the same time,* the whole ion was knocked into the faster rocking mode: 10 would become 01, the digits reversed. Using more such manipulations (there is no reason here to spell out the details), the 1 bit could be shuttled from ion to ion, traveling down the row like a bump in a Slinky—or a signal through a wire.

That was just the beginning of the possibilities. How an atom

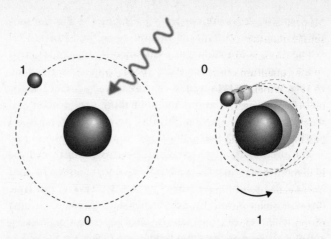

Figure 7.6. A quantum calculation. The atom's orbiting electron is in its high-energy state (1), while the whole atom is stationary (0). Hit with a pulse, the electron will drop down to the 0 state and set the atom to rocking: 10 becomes 01, reversing the order of the bits.

responded to a laser pulse could be made to depend on whether or not the row was rocking—just the kind of dependency needed to execute a logical operation. Hit with a pulse, an atom would flip to the opposite state, but only if the whole row of atoms was vibrating. The result would be the invisibly tiny equivalent of a controlled NOT gate, the building block of quantum computation.

This landmark paper took Artur Ekert's challenge another step toward completion. But it was still no more than a blueprint. The next goal was actually to build a gate from ions. Two researchers at NIST, David Wineland and Christopher Monroe, read a prepublication copy of Cirac and Zoller's scheme and decided to see if they could make it work. As part of the research on atomic timekeeping, NIST had assembled some of the best equipment available for isolating single ions. Wineland and

Monroe already had the makings of what turned out to be a simple quantum computer.

The two scientists were already experienced in showing that spooky quantum effects were not just the stuff of imagination. In 1986 Wineland belonged to one of the first teams to actually observe a quantum leap: An atom, hit with a photon of light, seemed to literally jump from a lower to a higher orbit without passing through the space in between.

A decade later he and Monroe joined several other colleagues in demonstrating an atomic version of a famous thought experiment called Schrödinger's Cat. To dramatize some of the paradoxes of quantum mechanics, Erwin Schrödinger whimsically proposed in 1935 that a feline be hidden inside a box along with a devious apparatus. Recall the two-slit experiment, in which a particle fired at a barrier has a 50-50 chance of passing through one of two holes. Now suppose that there are detectors at each of the two openings. If the particle goes through the top hole, nothing happens. It just sails on through. But if it goes through the bottom hole, a circuit is activated, releasing a hammer that breaks a vial of poison, killing the cat.

Schrödinger perversely suggested that until the box was opened and the outcome of the experiment observed, the particle must remain suspended in a superposition of two states: simultaneously going through both holes. But then, it seemed, the vial must also be in limbo, broken and not broken. And—the familiar punch line—the cat would be both dead and alive. To put it another way, the cat's fate is entangled with the particle's.

Recently many theorists have come to believe that they have found the way out of the Schrödinger mess. A single atom isolated from any outside disturbance indeed remains in superposition. But the apparatus in the cat experiment is constantly bombarded by air molecules and cosmic rays, subtly shaken by the vibrations from passing trucks, footsteps, minute seismic waves. These disturbances are all like little measurements. For that matter, the atoms that make up the cat and the box and the vial of poison, or any macroscopic object, are constantly inter-acting, "measuring" each other. Any superposition immediately collapses, ensuring that macroscopic objects like cats are here *or* there and not here *and* there, that they are either dead or alive. An isolated particle hovering in superposition is said to be in a state of "quantum coherence." Hence the process by which the tiny perturbations of the environment break down quantum ambiguity is called "decoherence."

As their stand-in for the cat, Wineland and Monroe used a single atom of a light steel-gray metal called beryllium, with one of its two outer electrons stripped away to create a positive charge. This made it possible to grip the ion inside a vacuum chamber, using as tweezers an electromagnetic field. The exper-imenters quieted the ion's frantic shaking with optical cooling, then used a laser pulse to put its remaining electron into a superposition: spinning both clockwise and counterclockwise.

For the finale, they used laser beams again, this time accom-plishing something that seems impossible: They carefully nudged apart the two halves of the superposition, separating them by about 100 billionths of a meter. The atom and its dop-pelgänger, one spinning this way, one spinning that way, momentarily seemed to exist in two places at once. Tiny distur-bances from the environment quickly caused the superposition to collapse. In later experiments the scientists timed how long it took for this decoherence to set in: about 25 to 50 millionths of a second. The farther the two atoms were pulled apart, the more vulnerable they became, and the faster the system broke down.

With their knack for suspending atoms in such delicate states, Wineland and Monroe thought they could probably make a controlled NOT gate. It might seem at first that this would require trapping at least two ions side by side, using one for the control signal and one for the NOT operation. But in the Cirac and Zoller scheme, two qubits of information can be car-ried by a single atom: It can be in a high-energy state or a low one, and it can be vibrating or not. For a classical analogy, think

of a switch that can be flipped up or down (one bit of information) or left or right (a second bit). Flip the switch diagonally,
into the 2 o'clock position, and it is up and right, 11, storing two
bits of information.

The NIST scientists began as usual by snaring and cooling a
beryllium ion. As in the Cirac-Zoller proposal, one of the qubits
was represented by the vibrational mode: whether or not it was
rocking. The second qubit was a little different: 1 or 0 was indicated not by the energy level of the outer electron but by its
spin. Still the basic idea was the same. The experimenters fine-
tuned the laser pulses so that they could reverse the direction of
the spin, flipping between clockwise and counterclockwise, but
only if the ion was vibrating in the 1 mode—the controlled
NOT operation.

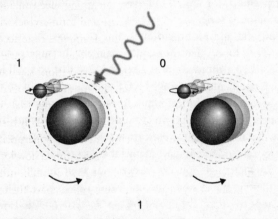

Figure 7.7. Another controlled NOT gate—one that was
actually made to work. When the atom is rocking (representing the 1 mode) the NOT function is turned on: strike
the electron with a laser pulse and the direction of its spin
will be reversed, from 1 to 0.

Finally, the experimenters needed a way to read out the result
of the calculation. Again they used a laser beam, shining it

inside the trap. By tuning the beam just so, they ensured that an ion in the 1 state would scatter back photons, signaling the digit with a tiny flash of ultraviolet light. A 0 ion would remain dark. In some of the very earliest digital computers, the result of a calculation was indicated by a numeral-shaped filament glowing inside a specially designed vacuum tube. History was repeating itself on a tinier scale. With a faint wink of light in an ion trap, the quantum processor was born.

In their paper, Cirac and Zoller estimated that, if all went well, an ion could be suspended in a trap long enough to do some serious computing. Then decoherence would set in. In less intrusive experiments, ion traps had been known to maintain superpositions for several minutes—a vast amount of time on the subatomic scale.

Wineland and Monroe didn't come close to achieving this ideal. They found that they could maintain the superposition for less than a thousandth of a second. But that was still long enough to execute a controlled NOT operation. Since then they have been refining the technique, forestalling decoherence a little longer. And they are trying to scale up the experiment. They know that if they hope to perform actual calculations, they will have to trap and manipulate many ions. Progress has been slow. The original single-ion experiment was done in 1995. Since then they have had some success manipulating four ions, and they hope to extend the procedure to several more. But it is difficult to refine the laser beam so that it flips just a single ion without also nudging its neighbors. And the more atoms that are strung together, the more they interact and the faster decoherence sets in.

They now believe that harnessing enough ions to make a rudimentary computer may require "multiplexing": interconnecting several ion traps to form an array of maybe dozens of processors. Each trap would hold a few ions. Responding to the beat of the laser guns, they would perform their simple computations, the output of one trap interacting with that of another, perhaps through fiber-optic lines. But that is way beyond the state of the art.

Those hoping to break codes with Shor's factoring algorithm, which would require thousands of qubits, might find the pace of quantum hardware development excruciatingly slow. The sophistication of the software far outstrips that of the machines on which it can be run. For solace one might look to the early decades of the twentieth century, when the first solid-state semiconductors existed only as "cat whisker" detectors mounted on homemade crystal-radio sets. By the mid 1940s, the first transistor emerged at Bell Laboratories, the beginning point of a technological arc that led from pocket-size portable radios to today's microprocessor chips.

With quantum computing the situation is a little different. There may not be a steady stream of interim technologies to help sustain the years of evolution. What is the quantum equivalent of the Japanese six-transistor radio? Still, experiments like the ones at NIST show that it is possible to make the quantum, single-atom version of the transistor (or maybe, more appropriately, the vacuum tube). With other labs also making incremental advances, there is reason to hope for a breakthrough.

8 | Counting with Atoms

At about the time Wineland and Monroe were tinkering with their ion contraption, other researchers were taking an entirely different approach to quantum computing. In the subatomic world, there are many different particles that can be used as quantum abacus beads. While experiments like those at NIST adopted atoms and their electrons as qubits, scientists in other labs have been trying to harness photons, the particles of light. Of course the ion-trap experiments also rely on photons, as the bullets to strike atoms and flip them between 1 and 0. This other research, centered at Caltech and at the École Normale Supérieure in Paris, goes further: it uses the photons to actually store the information. Here again 1 and 0 can be represented by whether a particle is in one state or the other. But there the similarities end. While the NIST scientists carry out their experiments in an ion trap, the other teams use an exotic technique called "cavity QED."

The initials stand for quantum electrodynamics, which is basically just quantum theory applied to photons and electrons. The cavity is a tiny chamber made of facing mirrors. Photons are captured between the reflecting surfaces, bouncing back and forth inside the narrow gap. During these repeated ricochets, the experimenters try to get the photons—"flying qubits," one scientist calls them—to carry out rudimentary computations.

The difficulty is that photons don't interact. Two flashlight beams meeting at an angle pass through each other unchanged. How then can light particles be made to flip each other's bits—to process information? The reason that the experiments work is that photons do interact with atoms. Struck by a photon of just the right frequency, an orbiting electron is excited to a higher energy level (the same thing that happens when a laser is used to knock one into another orbit, flipping between 0 and 1). When the electron relaxes back to its ground state, it expels a photon. So a photon goes in and a photon comes out. If the state of the photon can be altered in the process, a bit has been changed. By rigging up the optical equivalent of logic gates, the physicists can take one two-bit photon string and transform it into another, performing a computation.

In looking into any of this work, it is easy to become bogged down with details. With the flying qubits, for example, the frequencies of the light beams must be finely adjusted so that they resonate with the mirrored chamber and with the atom itself. (Think of bells tuned so that they pick up each other's rings.) Packed with such specifics, the papers describing quantum computing experiments are intimidating expanses of equations and graphs, with only an occasional oasis of prose.

Fortunately there is a limit to how much an outsider needs to know. From whatever direction you approach quantum computing, slicing machete-like through the undergrowth of detail, you arrive at the same destination. It is a matter of scientific taste whether experimenters work with ion traps, mirror-lined cavities, or even more obscure technologies. In every case, the underlying idea is the same: You trap some kind of particle and arbitrarily label two of its quantum states 1 and 0. You might pick spin, energy, mechanical vibration, charge—it doesn't really matter. The point is to get the particle to interact with another one that is labeled according to the same convention. Variables are tweaked so that the states of the particles emerging from the interaction depend in some well-defined way on the states of the particles going in. One pattern of 1s and 0s is converted into another, and that is what computation is.

In a classical computer, the nature of the counters is incidental: they can be silicon chips, vacuum tubes, electromechanical relays, marbles, billiard balls, Turing machine tapes, Geniac parts, or Tinkertoys. As long as an entity can be put into either of two arbitrarily defined states, 1 and 0, and as long as it can be linked so that its value is somehow dependent on its neighbors, the result is a computer switch. For quantum computation there is, of course, an amendment to be made: the counters can be 1, 0, or Φ, allowing myriad calculations to be performed simultaneously.

All the quantum computing technologies developed so far are exceedingly fragile. The slightest disturbance from the outside world can cause a bit to flip accidentally or force a superposition to collapse. This decoherence can occur a tiny fraction of a second after an experiment is under way. It is possible for a switch made from a single atom or particle to be flipped hundreds or even thousands of times before coherence breaks down. But that is hardly enough computing for the kind of earthshaking problem solving the visionaries have in mind.

Before complex programs can be run, more robust machinery must be made. And that means staving off decoherence—lengthening the window of opportunity for a calculation to unfold—while greatly accelerating the switching speeds. Both of these goals must be accomplished in a way that "scales," allowing hundreds and eventually thousands of qubits to interact, not just two or three.

Some encouraging strides have been made with yet another approach that involves a technology called "nuclear magnetic resonance," or NMR. Though it has its own weaknesses, the method allows for the manipulation of several qubits. And decoherence times are much longer—long enough to run simple algorithms consisting of dozens of operations, the first rudimentary quantum software.

The idea is to calculate using whole molecules. The choice seems natural. After all, the goal of quantum computing is to use long rows of quantum tokens to store and process information. A molecule is a ready-made chain of atoms, a qubit string. The electrons orbiting an atom's core, or nucleus, are not the only things that can be nudged between 1 and 0. NMR depends on the fact that the nucleus itself can be manipulated this way.

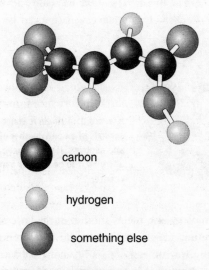

carbon

hydrogen

something else

Figure 8.1. A simple molecule. This string (primarily carbon and hydrogen atoms) can serve as a ready-made register for a quantum computer.

Though shown in diagrams as little spheres, nuclei are actually made from protons and neutrons, ranging in complexity from hydrogen, with a single proton, to unstable molecules like mendelevium (named for the inventor of the periodic table) with 101 protons and 157 neutrons. Like the more lightweight electrons and photons, these heavier particles have spin. Within a single nucleus the spins tend to offset each other. For every clockwise rotation there is a counterclockwise one to cancel it out. But if the nucleus is made from an odd number of the particles, there will be a spin left over. Thus the nucleus itself will have a net spin, 1 or 0.

These nuclear qubits are especially suited to store information. Surrounded by clouds of electrons, they are shielded from outside disturbances. In fact, the spins are so well protected that they can be held in superposition for entire seconds—eons compared with other techniques.

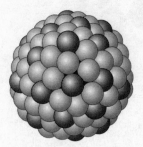

Figure 8.2. A single nucleus. If it consists of an odd number of protons and neutrons, it will have a net spin, allowing it to be used as a quantum counter.

The technology also meets another of the criteria for quantum computing: the ability to individually address each qubit, to flip it one way or another by hitting it with a well-timed pulse—consisting in this case of high-frequency radio waves. This is possible because of the NMR effect: Immersed in an intense magnetic field, a molecule's various atoms—hydrogen, carbon, fluorine, et cetera—will each respond to pulses of a different frequency. Even the same kind of atom—carbon, for example—may answer to a different call depending on its position within the molecule. A molecule is often described as a string of beads, but here it can be better thought of as a row of bells. With the high-frequency pulses scientists can reach into the molecule and "ring" a specific nucleus while leaving its neighbors alone. Its spin can be shifted between 1 and 0 or placed hovering in between.

To operate as a quantum computer, the nuclei also must be able to interact with each other, and this they do through their own magnetic fields. They are, in other words, entangled. Whether or not a nucleus flips in answer to a pulse from beyond depends on the states of its neighbors—whether they are aligned to say 1 or 0. And so they meet the basic requirement for doing logical operations. Finally, since the atoms in a molecule send out faint electromagnetic signals, the progress of a computation can be monitored with sensors and displayed as a pattern on a computer screen.

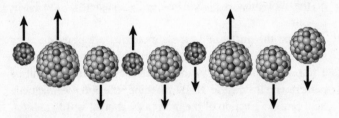

Figure 8.3. Nuclei lined up inside a molecule. Whether one flips in response to a pulse of radio waves depends on the states of its neighbors.

There is an additional factor about this approach that many researchers have found irresistible. The equipment for manipulating nuclear spins is already widespread. For years chemists have used expensive machines called NMR spectrometers to analyze chemical compounds. They surround them with magnetic fields and use high-frequency radio pulses to line up the spins of their nuclei. Then they read out the result on a computer monitor. (A similar technology, called magnetic resonance imaging, or MRI, is used by hospitals to scan organs and tissues.) As some physicists say, with a bit of hyperbole, people have been doing quantum computing all along. They just didn't realize it.

Since the early 1990s a number of labs—at Stanford, Los Alamos, the Massachusetts Institute of Technology, the IBM Almaden Research Center in California, the University of Oxford, and elsewhere—have been exploiting off-the-shelf NMR equipment to carry out simple quantum computations. The first step is to find, or synthesize, a molecule with the right properties. Suppose, to simplify somewhat, that it consists of a row of five atoms: ABCDE. These are the five qubits. (In a typical molecule there will be other nuclei as well, but they are not involved in the computation.) The substance is dissolved in a liquid, then a flask containing maybe 10^{21}, or a billion trillion, of the molecules is placed inside the magnets of the NMR machine.

At first the nuclei in all these molecules are pointing every which way—a random soup of spins, a blizzard of 1s and 0s. The result is a tabula rasa on which the computation will be performed. Inside the NMR machine, the intense magnetic field causes a fraction of the molecules, about 1 in 100 million, to line up so that their nuclei are all pointing the same way, 11111. This subset will be used to compute. Think of the molecules as 10 trillion quantum computers poised to carry out the same calculation. The result of this alignment is a characteristic signature—an electromagnetic signal consisting of all these identical voices—that can be detected against the sea of background noise and plotted as a pattern on the computer screen.

Sitting at a keyboard, the operator can pick one of the atoms, C perhaps, and ring it like a bell, flipping the 1 to 0. Throughout the soup, trillions of Cs will chime in synchrony. The strings now say 11011, and the pattern on the computer screen changes, verifying that the simple operation has been done.

Next the operator might choose to address the fifth atom, E. It so happens, let's say, that the molecule in the flask was origi-

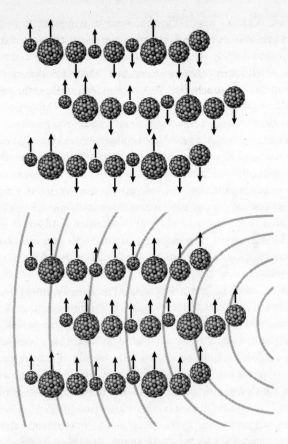

Figure 8.4. Molecular soup. In this solution of molecules,
the nuclei are initially oriented at random. Then an
intense magnetic field causes some of them to synchronize,
positioning them to compute.

nally chosen because its chemical structure ensures that E will
flip only if D is in the 1 state. Otherwise it will ignore its electromagnetic summons. This would constitute something that by
now is familiar: a controlled NOT gate, the building block from
which any quantum computer can be made.

This is clearly a very different kind of computer from the ones previously described. For one thing it consists of a liquid. More significantly, the same computation is being done in parallel by trillions of different molecules, what physicists call an "ensemble." That is mostly just an accident of the technique. NMR simply isn't precise enough to zero in on a single molecule. In any case, it takes all those molecular voices calling out together to produce a signal loud enough to detect.

Because of the robustness of the procedure and the number of qubits involved, NMR can be used not just for single operations, opening and closing a logic gate, but for carrying out long sequences of operations—computer programs. In 1999, researchers at Stanford and the IBM Almaden Research Center (including Lieven Vandersypen, Matthias Steffen, and Isaac Chuang) used a three-qubit molecular computer to run Grover's sorting algorithm. Their objective was modest: to search a "database" consisting of eight items—the binary numbers 0, 1, 10, 11, 100, 101, 110, and 111. By firing a barrage of 50 pulses at the molecule, the experimenters played the nuclei like piano keys, assembling on the fly the various chords of one- and two-qubit logic gates needed to carry out the precise steps of the computation. This involved a lot of subatomic manipulation. Retrieving one of the eight entries required two repetitions of the procedure—a total of 100 pulses.

A program that can find a needle in a haystack of eight straws is far from being a useful search engine, but the Stanford scientists went a little further. They were able to sustain the calculation for more than half a second, rapidly repeating it over and over before decoherence set in. Altogether they executed 280 tiny computational steps.

In a later experiment they synthesized a seven-atom molecule and used it to factor the number 15. (The answer, to no one's surprise, was 3 and 5.) Following Shor's algorithm, the machine churned through the steps, finding the period of a sequence of numbers using clock arithmetic and wave analysis—and it did so with exponentially fewer calculations than any classical apparatus could. There is no cause yet for cryptographers to worry. But it is now clearer that the kinds of operations needed to break industrial-strength codes can indeed be carried out subatomically.

The record for simultaneously marshaling the largest number of qubits is still 7, as of the year 2002. A 10-qubit machine seems to be on the horizon. But it is widely believed that it is unfeasible to move much beyond that with conventional NMR. Each time a nucleus is added to the molecular string, the electromagnetic voice of each qubit becomes relatively softer and harder to pick out from the din. The signal strength in fact fades exponentially. Special NMR machines might allow for quantum computers with perhaps 50 qubits—enough to carry out some interesting demonstrations and to model the behavior of small collections of subatomic particles. But that is still only a fraction of the number needed for a full-scale factoring machine.

There is no clear front-runner in this race. NMR has a much longer coherence time before breakdown than ion traps and cavity QED. But they have the advantage of faster switching speeds. All the methods used so far share an enormous drawback: Manipulating just a few qubits requires a laboratory full of equipment. (NMR machines generate their intense magnetic fields with superconducting coils bathed in liquid nitrogen and helium. Ion traps require a near-perfect vacuum as well as cryogenic temperatures.) It is difficult to envision quantum computing making its way into the world at large without someone developing the equivalent of the quantum microchip. Call it the Pentium Q: a solid-state device whose switches are individual atoms or particles.

The closest experimenters have come so far is an exotic technology called "quantum dots," little corrals of atoms on a chip in which single electrons can be confined. Each electron acts as a qubit. And the qubits can interact—electromagnetically or through a quantum phenomenon called tunneling—and perform logical operations. Some researchers believe that of all the technologies so far, quantum dots may be the easiest to scale, perhaps to a thousand qubits or beyond. But decoherence time is discouragingly rapid, and the experiments also have to be conducted at extremely frigid temperatures, near absolute zero. So far the effort has barely gotten off the ground.

Maybe it is just a lack of imagination that causes one to assume that future quantum computers must be built from something like today's silicon chips. A scientist at Bell Labs has proposed a device in which computing electrons would float on a surface of supercooled liquid helium. If a computer can be liquid instead of solid, then why not a calculating gas? The insides of a quantum computer may be nothing like what is normally thought of as a machine. Even if a quantum laptop turns out to be only a dream, a gargantuan quantum computer the size of a gymnasium and attended by a staff of hundreds would still cause a scientific upheaval—solving unsolvable problems, breaking unbreakable codes. Suppose the machine cost a billion dollars. It might be worth it considering the alternative: a rival government with the sudden ability to compute in a powerful new way.

Whatever solution emerges (barring the uninteresting possibility that nothing is found to work), physicists will be confronted with manipulating quantum tokens so tiny and delicate that the slightest disturbance can set a calculation askew. That raises the issue of how to deal with errors, the accidental bit flips that inevitably plague any computation and that are particularly troublesome in the quantum world.

With classical computers, errors are easy to identify and correct. The secret is redundancy. To ensure that a message is accurately transferred from point to point, within a computer or across a telephone line, send it three times. If one of the copies is different from the other two, it is probably defective. The majority rules. There is a slight possibility that, by chance, an identical bit flip will occur in two or even all three copies. So send the message five times or ten. As the amount of redundancy is increased, the likelihood of a conspiracy of coincidences shrinks toward nothingness.

There are more effective ways. If the message is 1011, render each bit in triplicate: 111 000 111 111. If a bit becomes flipped in transmission—111 000 101 111—it is easily noticeable and can be flipped back to restore the symmetry.

If there are two errors in the same threesome, the system breaks down. But double errors can be detected by encoding in quintuplicate: 1011 becomes 11111 00000 11111 11111. Now two bits can be scrambled in a cluster and corrected by majority vote. If the circuitry is extremely unreliable, each digit can be repeated many more times. This commonsensical observation lies at the heart of one of the most important intellectual achievements of the twentieth century, Claude Shannon's information theory: Any message can be transmitted with any degree of accuracy, as long as you're willing to add enough redundant bits.

Having to send three bits just so one good one gets through is a pretty clunky defense against error. A better way to use redundancy is to add markers called "parity bits" to a message. Say you're sending this string: 1011010. First, count the number of 1s. If the result is even, add a 0 to the end of the string; if it is odd, add a 1. In this case there are four 1s so the encoded version would be 10110100. When the packet is received, it is a simple matter to count the bits automatically and see if the answer coincides with the final bit. If it doesn't, one of the four message bits, or the parity bit itself, has been corrupted. Since there is no way to know which, the snippet of data must be resent.

That is the tradeoff: Only one extra bit is needed to protect seven. But finding an error means retransmitting the whole packet. When the probability of error is low and communication speeds are high, then the simple parity-check method works fine. Anyone who has used an old-fashioned (circa 1990s) terminal program to dial up another computer is at least vaguely familiar with this method. In the settings menu, you have to choose if you want "even parity" or "odd parity," and seven data bits or eight.

Fortunately there is a better way. By adding more than one parity bit, it is possible to close in on the precise location of an error. Then it can be corrected on the receiving end without requiring a retransmission. In a simple version of the "Hamming Code" (named for mathematician Richard Hamming), a

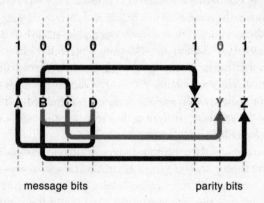

message bits   parity bits

ABC odd=1

BCD even=0

ABD odd=1

Figure 8.5. The Hamming error-correction code. Triplets of message bits are each entwined with a different parity bit. If a bit becomes corrupted, it is possible to zero in on the error and fix it.

four-bit message, call it ABCD, is followed by three parity bits, XYZ. These are each set to 1 or 0 depending on whether the parity of a certain cluster of bits in the message is odd or even. The first parity bit, X, is chosen to correspond to the first, second, and third bits of the message, ABC. The second and third parity bits, Y and Z, are likewise intertwined with two other triplets: BCD and ABD. Because the triplets overlap (and this is the main thing to remember), it is possible to deduce the location of an error by noting which parity bits are incorrect.

Try encoding the message 1000. There is a single 1 in the first triplet, an odd number, so the first parity bit is set to 1. The second triplet adds up to 0, which is an even number, so the second parity bit is 0. The third triplet, ABD, is odd, so the third parity bit is 1. The message with its redundant bits, 1000101, is then sent down the line.

If there is interference, the message might arrive looking like this: 1010101. The third bit has been flipped, but how can the receiver tell? The evenness or oddness of the three triplets is automatically recalculated and compared with each parity bit. It is immediately clear that the first parity bit is wrong. This clue raises the suspicion that an error occurred somewhere in the first cluster, ABC. The second parity bit is also off, so we have another clue: The error seems to be also among the digits in the second cluster, BCD. The clusters overlap, so we know that the error is either with B or C. The third parity bit is correct, so that indicates that the error was not with the third cluster, ABD. Thus B is exonerated and the culprit must be C. The erroneous 1 can be restored to 0.

By this process of elimination, one can always find which bit has been corrupted. Since the clusters overlap, an error in the message will always cause two or three of the parity checks to fail. (If only one parity check fails, then it is the parity bit itself that went astray.) To streamline the process, one can make a table showing which error is indicated by which pattern of incorrect parity bits. Of course in electronic communication the checking is done automatically.

CD players and other devices that must decode vast amounts of data use more elaborate schemes like the Reed-Solomon code (named for mathematicians Irving S. Reed and Gustave Solomon), which is based on something called "Galois field arithmetic." But no matter how complex the details, the basic idea is the same. The integrity of the message is protected by a phalanx of extra bits. Whether the value of these markers is set to 1 or 0 depends on various combinations of bits in the message. The result is a web of dependencies: If some bits are up, others must be down. If not, then something is wrong.

For a quantum computer, simple bit flips are just part of the problem. In a long string of qubits, the many computations are all taking place simultaneously. The slightest disturbance can cause information to leak from the system: the superposition collapses with Φs randomly becoming 1s or 0s. No matter how much progress scientists make in lengthening decoherence times or speeding up switching speeds, errors will occur, so there must be reliable methods to correct them.

The answer again is redundancy. But introducing this simple idea in a quantum system seems, at first, impossible. The classical error-correction schemes depend on looking for discrepancies between the patterns of 1s and 0s in the message and the redundant parity bits. But that involves determining the value of the bits—measuring them—and in a quantum system a measurement leads to an abrupt collapse. Quantum computers would have to find and repair errors without actually reading the qubits. This seeming paradox was used for years as an argument against the possibility of quantum computing: In such a fragile system, errors would constantly occur, but trying to correct them, it seemed, would make things worse, creating more decoherence.

A breakthrough came in the early 1990s when researchers showed that quantum error correction was theoretically possible—an idea improved upon over the next two years by an Oxford scientist, Andrew Steane, and Peter Shor, the inventor of the factoring algorithm. The scientists showed, quite unexpectedly, that the peculiarities of quantum mechanics make it possible to fix errors without actually reading and upsetting the offending qubits. The key again is entanglement, the ability to correlate quantum particles so that their fates are intertwined. If one is up, the other must be down. Entanglement is in itself a kind of quantum redundancy. It makes it possible to determine that a qubit is in error and restore it without prematurely learning what it says. If no information leaves the system, there has been no measurement and the computation remains intact.

Of all the ideas described here so far, quantum error correction comes closest to straining the limits of what is possible to convey with words. (As one scientist unhelpfully explains, quantum error correction "is essentially a way of embedding one finite-dimensional Hilbert space inside another larger [higher-dimensional] Hilbert space.") But it is possible to distill some of the flavor of the idea with an imperfect analogy.

A Hamming code can be thought of as involving a classical version of entanglement. The message bits are intertwined with the parity bits. If some are up (1), others must be down (0). By studying these relationships, errors are identified and corrected. But go a little further and imagine a code in which the two groups of bits are somehow linked dynamically—within the message. If a bit is flipped during transmission, some internal mechanism will actually cause the corresponding parity bits to flip with it, restoring the proper correlation. Say we are sending 1000101—four message bits and three parity bits. If the fourth bit is flipped, so 1000 becomes 1001, then the parity bits are instantly updated from 101 to 110. The message itself, flying through the wire, works like a little machine.

Now take the idea another step. Surround the four-bit message with *two* groups of parity bits, one on each side. The right-hand group is static: It acts as a memory register, displaying the parity of the original message as it was transmitted. The left-hand group is dynamic—linked to the message bits, readjusting itself in flight to keep track of any changes. When the message arrives, the two clusters of parity bits are compared. (In fact, we can imagine that the message contains even more internal links, allowing it to make the comparison itself.)

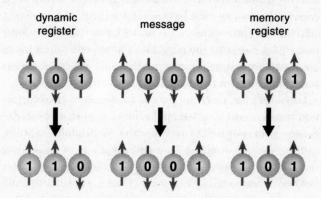

Figure 8.6. A message that corrects itself in flight. A scrap of quantum information, "1000," represented perhaps by a string of spinning particles, becomes scrambled during transmission, the last digit changing from 0 to 1. But since the message bits are entangled with those in the left-hand register, its value also changes. The receiver can compare it with the right-hand register and see that it no longer matches.

If 1000 has mutated to 1001, the two parity clusters will no longer jibe, and the mismatch instantly shows that the problem must be with the fourth message bit. Note that it is not actually necessary to look at this bit and determine whether it is 1 or 0. This is a binary system, so if we know the bit is wrong we simply need to reverse it, whatever its value. We can close our eyes and flip it back without having ever read the message. The bits read themselves and the information remains within the system.

There is no obvious way to take classical bits and tie them

together so they interact in flight. But qubits are inherently more powerful. They can be entangled quantum-mechanically (remember again the spooky EPR effect), so that a mechanism something like the one just described is built into the message. Qubits can be linked so that they affect one another on the fly.

With entanglement—quantum redundancy—a qubit's value is, in other words, spread among several qubits. If one becomes corrupted, the values of the others will be affected in a very specific way. To check the integrity of a cluster, you can read the values of the extra qubits and deduce where there has been an error, *but not what the error is.* (That would constitute a measurement.) Then, without knowing the offending qubit's original value, you can nudge it with a properly timed pulse that sets it to the opposite value: 1 if 0, or 0 if 1. The qubit itself has not been read, only its protectors. Thus an error can be located and repaired without information leaving the quantum realm.

This kind of error correction requires redundancy with a vengeance. The original schemes used as many as 9 qubits to record a single 1, 0, or Φ. Since then the number has been reduced to a minimum of 5, but that still would mean quintupling the number of atoms or other quantum registers—just to take care of single-bit errors. What if 2 or 3 or even more of the qubits in a cluster went askew? To protect against multiple errors, you would need to surround each qubit with dozens of redundant qubits. And then, when an error was found, you would have to worry about more errors creeping into the correction process itself. Guarding against that would require still more redundancy.

The situation may not be hopeless. Calculations show that with enough extra qubits, one could make a computer accurate enough to factor long numbers and break codes. For a number hundreds of digits long—the kind that could take eons on a classical computer—a quantum computer might need thousands or even tens of thousands of qubits.

With the current state of the art, that almost seems absurd.

But imagine that we are back in the 1940s, when a room-sized supercomputer like Eniac consisted of thousands of fragile vacuum tubes, which constantly burned out and had to be replaced. What if someone had proposed that to perform really great calculational feats—processing so much data so rapidly that you could render full-color moving pictures on a computer screen—you would need a processor with millions of vacuum tubes and a memory bank consisting of a billion more . . . and that, considering how unreliable vacuum tubes are, you would need to multiply that number to add error-correcting redundancy. Almost anyone would have thought this was mad. Yet that is just the contraption that millions of people have on their desks and even their laps: a personal computer. The only fundamental difference is that the vacuum tubes have been replaced by microscopic transistors etched in silicon.

A few years after quantum error correction was discovered, Manny Knill and Raymond Laflamme at Los Alamos actually used NMR to run a simple scheme in which a qubit was entangled with four others (all within the same molecule). An error was purposefully introduced, identified, and corrected—all by sending the properly crafted sequence of pulses.

For all the convolutions needed to correct errors, scientists have shown that with enough redundancy it should be possible to carry out arbitrarily long quantum calculations. At least nothing in the laws of physics seems to rule it out. The more complex the algorithm, the more extra qubits are required, but the number, to everyone's relief, does not seem to explode exponentially. No one knows which technology might allow the large-scale harnessing of quantum bits that would be required. But the search is in its earliest stages. Everything in the universe is made from quantum particles, from potential qubits. There are so many possibilities yet to try.

9 | Quantum Secrecy

Long before quantum code breaking becomes a reality, physicists working in another corner of the field may have developed the perfect antidote: a means of using quantum mechanics to make codes that are undecipherable, as guaranteed by the laws of physics.

Even in the classical realm it is possible to make an unbreakable code. Just turn the message into one long number and add it to a random number of equal length. If this key is truly (not pseudo-) random—generated, for example, by decaying atomic nuclei—then no one but the sender and receiver can decode the message. There will be no pattern to uncover, no doorway inside. Unless the key is intercepted. That, of course, is the problem: There is no foolproof way to protect the key. For now, the near impossibility of factoring long numbers all but ensures the safe passage of information. But with quantum computing, the layers of protection would be stripped away.

What if, in a kind of mathematical jujitsu, quantum mechanics could be pitted against quantum mechanics, fighting fire with fire? We've already seen hints of how this might be done. Quantum information cannot be read without scrambling it. Attempts at eavesdropping would be immediately apparent. And other quantum effects—like the "spooky action at a distance" manifested by the EPR effect—suggest the possibility of somehow linking senders and receivers in subtle ways that go beyond what can be done with classical physics.

Putting these ideas together, physicists have devised a means of producing and distributing a secret key that is impossible to intercept. Using quantum mechanics, a sender and receiver can simultaneously generate the same random number, consisting of a long string of qubits, in two different locations, *without the number itself actually traveling in between*. Furthermore, any attempt to listen in on the transaction is immediately apparent. Experimenters have already distributed quantum keys over distances of several miles. Many people in the field believe that this kind of communication will give quantum information its first niche in the practical world.

Quantum cryptography has its roots in what originally seemed like nothing more than a bizarre thought experiment: a quantum anticounterfeiting scheme. For centuries governments have used watermarks, special inks, and tiny threads to make it extremely difficult to faithfully copy paper money. With enough resources and ingenuity, these protections can be circumvented. But it is impossible to copy quantum information. To do so you must measure it, breaking down the superposition. So why not mint money with qubits for serial numbers? Just trying to read the quantum signature would render the bill invalid. The counterfeiter would not only fail to make an exact copy; in the attempt he would destroy the original.

The idea was stumbled upon, in the late 1960s, by a Columbia University graduate student named Stephen Wiesner. Each qubit, he imagined, would be represented by a photon. It is a familiar idea by now that a particle can spin up (counterclockwise) or down (clockwise). With photons there is another possibility: They can also spin sideways, with the axis horizontal.

Another way to say this is that light can be aligned, or "polarized," either vertically or horizontally. Polaroid sunglasses

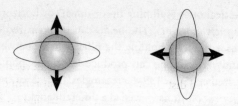

Figure 9.1. Spinning photons, vertical and horizontal

exploit this phenomenon to reduce glare. Most light reflected off the surface of a road or the hood of a car is, like its source, aligned horizontally. So the lenses act as a filter (think of them as gratings of finely spaced vertical bands) that discourages the passage of horizontal light. If a lens is turned 90 degrees, the filter will be pointing the wrong way, and the glare will shine through. Hold two of the lenses face-to-face and turn one so that they are aligned perpendicularly. Now both directions are blocked and the image will become nearly opaque.

Viewed at the finest grain, light is made of photons. By holding a polarizing filter in the path of a laser beam and twisting it into different positions, one can produce "rectilinear" photons,

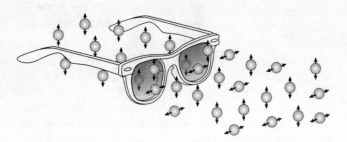

Figure 9.2. A polarizing filter. Only photons oriented in a certain direction can pass through.

either vertical ones (spinning up or down) or horizontal ones (spinning left or right). The beam can be polarized in other directions as well. By holding the filter halfway between horizontal and vertical, one can produce "diagonal" photons that are polarized at 45- or 135-degree angles. (These, of course, are just rectilinear photons viewed at a different angle.)

Let us decree that | and \ photons represent the binary digit 1, while — and / represent 0. (This is a little different from previous encoding schemes, but the general idea is the same.) Since these are qubits, it is not surprising that they can be intertwined in unusual ways. Heisenberg's uncertainty principle guarantees that certain quantities like position and momentum are mutually exclusive—the more precisely you know one, the fuzzier your knowledge of the other becomes. This complementarity, as it is called, also applies to the rectilinear and diagonal polarization of light. Precisely measuring one destroys all knowledge about the other. This truth forms the basis of the anticounterfeiting proposal.

Suppose you have a photon that has been oriented vertically or horizontally. Send it through a cross-shaped rectilinear filter, resembling a plus sign (+), and it will emerge unscathed.

Figure 9.3. A filter shaped like a cross. Either vertical or horizontal photons pass right through.

But now try sending a diagonally oriented photon. An arrow tilted at a 45-degree angle can be thought of as being halfway between horizontal and vertical. It is, in other words, a superposition: 50 percent one way and 50 percent the other. Try to squeeze the particle through a rectilinear filter and the superposition will be destroyed, forced to become horizontal or vertical. And, as we've seen, such decisions must be made at random. The orientation will now be horizontal *or* vertical, and in the process information about whether the photon was originally at 45 or 135 degrees is erased. Measuring rectilinearity precludes measuring diagonality—Heisenberg again.

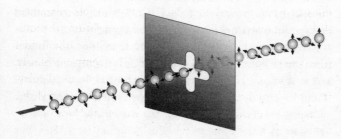

Figure 9.4. Diagonal photons are forced to assume the vertical or horizontal position. The choice is made at random.

To make a bill tamperproof, it would be marked somehow (remember, this is just a thought experiment) with a row of photons, some aligned rectilinearly and some diagonally. To copy this unique signature, the counterfeiter must measure each qubit, one by one, to determine its orientation. Since he has no way of knowing which way the particle is polarized, he has to guess. Say he chooses to hold the filter in the rectilinear position. If the photon is indeed aligned that way, it will pass through the filter unaltered. But if the counterfeiter is wrong and the photon is diagonal, Heisenberg's principle will kick in. There is a 50-50 chance that the filter will force the photon into becoming vertical or horizontal. Either way, the knowledge is useless. Information about how the photon was originally pointing is gone. On average, half the measurements will be incorrect, and the counterfeiter has no way of knowing which ones.

That is the beauty of the scheme. Suppose he tried to spend the bill with the miscopied number. In this plan, money is also marked with an ordinary serial number, printed in old-fashioned ink. By referring to a master list matching these with the corresponding qubit patterns, a bank could quickly tell a merchant whether a bill was genuine. (The bank's measurement also wipes out the number, so before the bill is released, the qubits must be reset.) In fact, if the counterfeiter tries to spend the original bill, he will also be out of luck, for he has scrambled its quantum signature in the attempt to duplicate it.

No one thought it would actually be feasible to mint quantum money. Photons can be stored for only the briefest instant, and that requires temperatures near absolute zero. Each bill would have to be transported in a refrigerated van containing an expensive cryogenics lab. But the theory itself was sound. The laws of finance might prohibit quantum banknotes, but not the laws of physics.

Wiesner passed the idea on to his former college roommate, Charles Bennett, who is now at the IBM Thomas J. Watson Research Center in Yorktown Heights, New York. Years later, at a meeting in Puerto Rico, Bennett mentioned it to Gilles Brassard, a computer scientist and cryptographer at the University of Montreal. Fascinated by the possibilities, they began thinking about whether quantum principles might be used to protect coded information against eavesdropping. If information could be sent through some kind of quantum channel, any attempt to intercept it would cause random disturbances easily detectable on the other end. Before long they were collaborating on a detailed plan.

Because of its optical properties, a crystal of calcite, a transparent substance, can be used as a photon sieve. Align the crystal horizontally and send a stream of photons in its direction.

Those with the same alignment will pass through. Those that
are vertically aligned will be slightly offset.

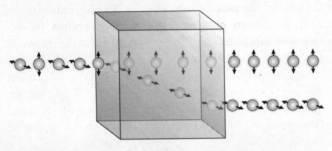

**Figure 9.5. A photon sieve. The crystal sorts horizontal and vertical
photons into two different streams.**

But suppose you take the photon source and twist it 45
degrees. Now you are sending diagonally polarized photons
instead of rectilinear ones. Since a diagonal photon is halfway
between vertical and horizontal—in a superposition—it will be
forced to choose between these two directions. It randomly
emerges as horizontal or vertical and the original information,
whether it was leaning left or right, is lost. A thought experiment has been brought to life. Rectilinear photons are properly
registered, as horizontal or vertical, but diagonal photons are
randomized.

Now picture the converse situation, twisting the crystal 45
degrees. This time the diagonal photons will be properly measured, as / or \, and the rectilinear ones will be scrambled. A
horizontal or vertical line is a superposition of diagonals.

Bennett and Brassard realized they could use this idea to
securely distribute a cryptographic key. In fact, the key would
not actually be distributed but would be created simultaneously
at both ends of the transmission.

When cryptographers speak of sending a message from point

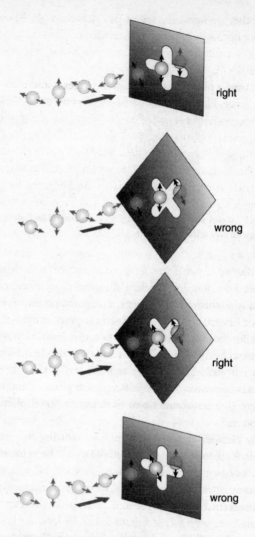

right

wrong

right

wrong

Figure 9.6. As Bob receives his stream of photons, he guesses how to hold the filter. When he is right, he will get a valid measurement. When he is wrong, the photon will be forced into an incorrect position.

A to B, they traditionally call them Alice and Bob. To create a key, Alice begins by sending Bob a stream of photons, randomly twisting her filter so that some are rectilinear and some are diagonal: | − / \.

Each time a photon arrives, Bob measures it and records the results. Since, like the counterfeiter, he has no way of knowing which way to hold his own filter, he must guess. Suppose that the first photon is polarized horizontally and he happens to use the rectilinear filter. He will then have the correct orientation. If the photon is vertical, he will also get the right answer. But if the particle is oriented 45 degrees or 135 degrees, he will be wedging it through the wrong sieve. The measurement will be bungled. When Alice's transmission is done, Bob ends up with a list of measurements. But he does not yet know which are good and which are bad—how many times he aligned the filter correctly.

He could call up Alice and ask how each photon had been originally aligned, but that would defeat the purpose. An eavesdropper (inevitably called Eve in the literature) might be listening in. But there is another way. Bob contacts Alice by phone or e-mail and, without revealing the actual results of his measurements, tells her which filter he used for each photon. Photon 1, he says, was measured using the rectilinear filter. That is right, she says. Bob now knows that the result of his measurement, horizontal, is correct. He doesn't need to tell Alice this. She already knows because it was she who sent the photon in the first place. Alice and Bob each put a check mark on their lists by the first photon and move on. For the second photon, Bob says that he used the diagonal filter. Wrong, Alice replies. Whatever measurement he got will be incorrect, so they both cross out photon 2.

When the process is done, the bad measurements are discarded and Alice and Bob are left with the same random string of information. This they translate into binary: | and \ marks mean 1, and – and / marks mean 0. They now have the key that can be used to encrypt the message. Alice takes the text (also rendered in binary form) and adds the key to it. Then she transmits the result however she likes, through any old classical channel. She could print it in a newspaper or broadcast it on TV. At this stage, security is irrelevant. Only Bob can know the proper number to subtract.

What if the eavesdropper, Eve, intercepts Alice's original transmission of photons, tapping somehow into the quantum channel? She surreptitiously measures each qubit then passes it along to Bob, so quickly that he doesn't notice the delay. Then she waits for him to do his own measurements and listens in when he and Alice determine which of the bits to use. The key, it seems, would be created in three places.

But quantum physics prevents this. Like Bob, Eve cannot know which filter to use. If she picked the rectilinear one for a photon that was horizontal or vertical, she would acquire accurate information. She could disguise her interloping by sending on a duplicate with the same specs. But she is just as likely to pick the diagonal filter, so half the time her data will be wrong. Instead of passing along a diagonal photon she would unwittingly send one that was either vertical or horizontal—a discrepancy that Alice and Bob could easily detect.

To test whether their data have been compromised they can pick a sample of their good bits (the ones for which they both used the same filter) and compare results over the phone. If they have the same readings, they can be confident that no one has intervened. These bits are then discarded (in case Eve was listening) and the remaining ones are used for the key. But if some of their answers are different, someone may have been interfering. They can discard all the data and try again.

In 1989, on the tenth anniversary of their meeting in Puerto Rico, Bennett and Brassard moved beyond speculation and performed the first successful quantum cryptographic transmission. Using light-emitting diodes and devices called Pockels cells—think of them as polarizing filters that can be automatically rotated with electronic signals—they sent a message across a tabletop, a distance of about thirty centimeters, or one foot. Thus Alice and Bob (who were represented by simulations on a personal computer) were able to agree on a secret key.

In later experiments in Switzerland, scientists have sent qubits through fiber-optic telephone cables between Geneva and Lausanne. The technique has recently been commercialized. Using two desktop devices (called Alice and Bob), customers of a company called id Quantique can exchange quantum keys over a distance of about seventy kilometers or about forty-three miles. (And if you don't want to fool with cables, be patient. In New Mexico, Los Alamos scientists have exchanged a quantum key over a mile of open air.)

Experiments like these are still little more than proofs of principle. It is disappointing to read the fine print on the papers: So far, it has not actually been possible to consistently send qubits a single photon at a time. The technology is not that refined. Instead the scientists use as their qubits dim flashes of light. This works pretty well, but it opens up an opportunity for eavesdropping. Suppose (reverting to the realm of the thought experiment) Eve inserts a half-silvered mirror into Bob and Alice's photon stream, diverting part of each flash. If a bit is made from multiple photons, all with the same polarization, she would be able to skim off a few, saving them until she learns which filter to use. Bob and Alice could make her task harder by making the flashes as dim as possible. But the fainter they are, the shorter the distance they can be sent.

This is not a fatal flaw. It could be very useful to send quantumly encrypted data within a single building or even within a single device like an automatic teller machine. And longer ranges could be divided into many shorter hops. When sending classical data—bits rather than qubits—through fiber-optic lines, the pulses are boosted along the way by electronic devices called repeaters. That won't work here, for it means repeatedly measuring and copying the signal, and you cannot do that with quantum data. But you could read and then reencrypt the message at every transfer station. Picture a chain of Bobs and Carols and Teds and Alices collaborating in a cryptographic relay race. As the fading message arrived at one station, it would be read and then reencoded, going from quantum to classical and back to quantum, and so forth down the line.

In what has been described so far, Bob and Alice use their photons as soon as they are sent, measuring them to make a random key. But it might be convenient, particularly in the case of quantum relay stations, to be able to store a supply of polarized photons, taking them from cold storage only as they are needed for encrypting a message. After Bennett and Brassard invented their protocol, Artur Ekert, the Oxford theorist, devised an alternative scheme, one based on the EPR effect. Two photons shooting in opposite directions from an atom are each in superposition, spinning two ways at the same time, and they are correlated quantum-mechanically. The moment one is measured, it snaps into focus, spinning one way or the other. And its partner, no matter how far away, instantly assumes the opposite orientation.

So suppose Alice generates pairs of entangled photons, keeping one for herself and sending the other on to Bob. Each party keeps its supply stored until needed. When they want to exchange a message, they remove photons from their cryogenic hoppers, one by one, and make various measurements, comparing results over the phone. As in the Bennett and Brassard scheme, they tell each other which way they held their filters but not the results of the measurements. Since each pair of photons is entangled, Bob's horizontals are Alice's verticals, and vice versa. Knowing this they generate their random key.

Now consider the instances in which they hold their filters at different angles, say 30 degrees apart or 45. Analyzing these mis-measurements reveals whether Eve was listening in. Without getting into details (which I have relegated to the notes section), consider that a well-known result called Bell's Inequality proves that pairs of entangled particles are more strongly correlated than any two classical objects can ever be. This is very difficult to understand, much less explain, but the upshot is this: When Alice and Bob hold their filters at different angles, sometimes their measurements will match (Bob horizontal, Alice vertical) and sometimes they will not. But for certain misalignments, their results will coincide *more often than is possible in a world ruled entirely by classical physics.* Hence, if someone tried to measure Bob's photons while they were in flight, the quantum purity would be corrupted and the correlation would revert to the weaker classical one.

In theory, the EPR effect can also be used to carry out what physicists call "quantum teleportation." The various attributes of a particle—mass, spin, and so forth—can be copied and transferred to another particle, creating an exact replica. In the process, the original is destroyed. Though far from what Gene Roddenberry envisioned in *Star Trek,* quantum teleportation came as a surprise to theorists. After all, it is supposed to be impossible to read quantum information without destroying it. But (skimming over the details), the EPR effect can be used to ferry the information from one particle to another without actually measuring it. (The flavor of the technique is similar to quantum error correction, where corrupted bits are detected without actually being read.) Finally, through a classical chan-nel (a phone call or whatever), the sender can provide addi-tional data needed to re-create the original.

Though closer to realization than a working quantum com-puter, practical quantum cryptography remains in an early stage. In the thought experiments, theorists blithely speak of generating and manipulating photons one at a time, shooting them through space, retrieving them unscathed, and storing them for as long as desired. All this is even trickier than it sounds. But small milestones continue to be made. Scientists at the University of Cambridge and at Toshiba have made light-emitting diodes that expel one photon at a time. But the process only works at very cold temperatures. In another development, scientists at the Massachusetts Institute of Technology, the United States Air Force, and two other laboratories recently figured out how to freeze light to a standstill, briefly storing the photons inside a supercooled crystal.

There is still a long way to go. In the meantime, by increasing the size of the key, classical cryptography can be made as secure as desired—until someone comes up with a working quantum computer.

The Hardest Problem in the Universe

Every second throughout the biosphere molecules called proteins effortlessly solve a problem that has stumped biologists for many years. A single protein is strung together from hundreds of chemical beads called amino acids. But before it can do anything, the long chain must spontaneously bunch up into the unique three-dimensional configuration that determines its purpose in life. Some proteins, like actin and tubulin, are structural, forming the scaffolding of living cells. Others work like nanoscopic machines. Chlorophyll converts light into sugar to feed plants; hemoglobin absorbs oxygen and expels carbon dioxide like an invisibly tiny lung—"nature's robots," some biochemists call them.

The mystery is how a newly minted protein "knows" how to fold up, finding its way through the labyrinth of possibilities and settling into the correct shape. Think of all the ways there are to crumple a sheet of paper. Then imagine having to do it the same way every time.

For a typical protein, the process begins when it comes off the cellular assembly line. First it twists itself into a long flexible coil like a telephone cord. Then this writhing helix begins to bunch up into a more complex configuration. Atoms with similar charges push apart, atoms with opposite charges pull closer

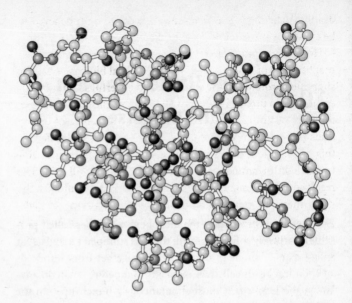

Figure 10.1. A complex protein. How does it "know" how to fold up the right way?

together. Some of the amino acid beads are "hydrophobic," or water-fearing, avoiding the cellular fluids by congregating inside the molecule. Others are hydrophilic, migrating to the outer edge of the protein.

There are so many ways this three-dimensional tug-of-war can play out that seemingly it should take eons to find the path that leads to the correct form. Like exhaustively searching for a route through a labyrinth, searching through the space of possible foldings suffers from the malady called "exponential explosion." Yet, oblivious to the mathematical arguments, chains consisting of thousands of amino acids fold up in minutes, shorter ones in a fraction of a second. Somehow they are able to avoid the many twists and turns that can lead down a wrong avenue to a misshapen structure—one that is impotent or even deadly. Alzheimer's and mad-cow disease are both believed to be caused by protein folding gone astray.

Hoping to understand this precise molecular gymnastics, practitioners of a field called "computational biology" simulate simplified proteins on the world's most powerful supercomputers and try to predict how they will fold. Somehow, they believe, the final outcome must be implicit in the sequence of amino acids that makes up the molecular chain. If they could decode this information, they would be able to start with some arbitrary chain of amino acids and predict how it will fold. The result could be designer proteins that would bunch into just the right shape to perform some biological function—a high-precision drug. But for many researchers, the main appeal is purely intellectual. Can they outwit the molecules at their own game?

Every two years they hold a competition. Some of the simulations do better than others, but no one has come close to mimicking what proteins can do with ease. Somehow they are able to solve almost immediately a problem that appears to be computationally intractable. Just what this means opens up a world of philosophical debate.

Protein folding is a member of a class of problems that has become notorious in the mathematical world. There are harder ones (and we will hear about some of them later), but for many mathematicians and computer scientists these are the most interesting. Other examples include exploring a maze by trial and error, putting together a jigsaw puzzle, finding the most efficient way to pack shapes into a box, or, most famously, solving the traveling salesman problem: Given a list of cities, find the shortest route a traveler can use to visit each one, with no backtracking allowed.

As with protein folding, one quickly comes up against a proliferation of possibilities. Say there are just 10 cities on the itinerary. Whichever one you decide to start with, you are then

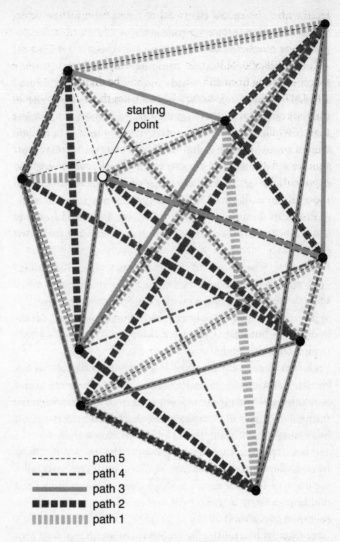

starting
point

---------- path 5
- - - - - path 4
━━━━━ path 3
▮▮▮▮▮ path 2
||||||| path 1

Figure 10.2. The traveling salesman problem. Here are a few of the 1,814,400 different ways to connect 10 points. Which is the shortest?

faced with 9 choices for city number 2, and from each of those, you must decide among 8 possibilities for city number 3. Altogether the number of routes is $10 \times 9 \times 8 \times 7 \times 6 \times 5 \times 4 \times 3 \times 2 \times 1$, or 3,628,800. Since that figure actually counts each route twice—starting from either end—it must be divided by 2 to get 1,814,400 unique trajectories. For 11 cities there are almost 20 million routes, for 12 there are almost 240 million, for 15 more than 650 billion, and for 20 more than a quintillion (a million trillion or 10^{18}). As the number of destinations increases, the time to sift through all the paths to find the shortest grows exponentially—as it does for the protein-folding problem as more amino acids are added to the chain.

Problems like this are called "NP-complete." The initials stand for "nondeterministic polynomial-time." The name is not immediately revealing. Easy problems belong to the class P, meaning they can be solved with relative ease, in what is called "polynomial time." (As the problem grows in size, the time it takes to solve it increases at a comparatively gradual pace.)

The NP-complete problems could also be solved in polynomial time—but only if you were equipped with an imaginary device called a nondeterministic computer. At each step of a problem, this omniscient machine would randomly decide which of the many computational paths to choose—and it would be 100 percent lucky, always guessing the right one. If one of these electronic oracles could be built, then the traveling salesman problem and its equivalents would be trivial.

Alas, since real computers are deterministic, there seems to be no foolproof way to solve NP-complete problems in less than exponential time. Of course it is possible that you will be lucky and be given a particularly simple instance of, say, the traveling salesman problem. If all the cities lie on a circle or in a straight line, finding the most efficient route is easy. But a general method for solving any traveling salesman problem has so far proven computationally elusive. It is possible that there is an undiscovered algorithm lurking in Plato's heaven that would allow ordinary computers to rapidly find the answer. No one has proved that there isn't. But as the decades pass, mathematicians have become resigned to the likelihood that NP-complete problems will remain beyond their grasp.

Things could be worse. A more comforting characteristic of these problems is that, hard as they are to solve, once you have a solution it can be easily verified—in polynomial time. When you begin a jigsaw puzzle you are faced with an empty expanse, a pile of pieces, and an overwhelming number of juxtapositions to try. But once you have blundered through to a solution, you can see at a glance that it is correct. In the case of the traveling salesman, you would have to restate the problem slightly: Is there a path shorter than x that will visit each city once and only once? If some oracle with access to a magic means of calculating, or a hell of a lot of time, reveals an answer, it is easy enough to measure and see if the route indeed meets the criteria.

Faced with a problem that is plagued by exponential explosion, it is natural to wonder whether quantum computation could provide the key. After all, quantum mechanics, with its irreducible randomness, is nondeterministic to the core. Could a string of atoms somehow be programmed to act as a nondeterministic computer, solving these problems? If so, this would be among the greatest mathematical achievements of all time.

Another thing that mathematicians have proved is that all the NP-complete problems are connected. Solve one and you can solve them all. Learn to find the optimal route for the traveling salesman and you can predict how proteins will fold. If this were to happen, the world would become a very different place, and not just for itinerant peddlers. Thousands of problems in many different fields have been shown to be NP-complete. Laying out the most efficient path of wires, whether for a computer chip a few centimeters across or a communications network that stretches for miles, is a version of the traveling salesman problem. Closely related are the problems of efficiently scheduling airline flights or finding the optimal order in which to perform a series of tasks in a production line. In many cases mathematicians have devised fairly good ways of approximating an answer—one that is satisfactory if not ideal. Sometimes, though, an exact solution would be desirable, if only it were possible to obtain.

Consider software design. Whenever a new program like Word or the latest version of an operating system like Mac OS X or Windows XP emerges, it must go through an extensive period of testing for bugs. With millions of lines of computer code some are bound to clash. Will hitting Delete on the word processor while iTunes is playing cause the system to crash if Eudora is simultaneously checking for mail? Or how about pressing Shift + Delete and double-clicking in a window's title bar—or triple-clicking, or any of the exponentially exploding number of combinations of commands?

Mathematicians call this a "satisfiability problem." The classic example is planning a party. A will only attend if C does and E doesn't, while C requires that B and G be there. G, however, will not be in the same room with A. How can you satisfy them all? Send out 10 invitations and you're faced with 2^{10} or 1,024 combinations to consider. For 20 the figure surpasses a million, for 30 a billion, and for 40 a trillion.

Satisfiability problems are NP-complete, which is why software is sold with no guarantee. The only way to really know what will happen at the party is to wait and see. The best you can do with software is to farm it out to an army of so-called beta testers and try to fix the problems they stumble upon. But some bugs will inevitably be left for the buyer to find. What you are purchasing remains a work in progress, a truth that is particularly disturbing considering how much of our civilization, including its financial and defense systems, relies on software code whose consistency can only be tested by trial and error.

If there were a way to definitively solve NP-complete problems, then Apple or Microsoft could run a new piece of software through a verification machine. With a few slices of its algorithmic scalpel, it would cut through the tangle of possibilities and reveal whether a program suffered from internal inconsistencies.

The same could be true of any mathematical or logical system. The basic layer of software in a society is its legal system, a complex web of constraints on human behavior that has accreted over hundreds of years. Conflicts and inconsistencies—bugs in the program—are unavoidable. Many are expunged during the never-ending beta testing of court cases and legislative debates, but new ones are constantly being introduced.

Suppose a society decided that, at any cost, it wanted an internally consistent legal system. First it would have to launch a mammoth project to translate every provision of every law into a simple logical statement. And that would be just the beginning. Each statement would have to be checked for consistency with all the others. The fifth statement would have to mesh with statements 1 through 4—and with their various combinations: 1 with 2, 1 with 3, and 1 with 4 . . . and 2 with 3, 2 with 4, and 3 with 4. And with 1, 2, and 3 combined, 1, 3, and 4, and so on. As statements were added to the pool, the difficulty would grow exponentially. Proving the consistency of even a small section of obscure municipal code would be a computational nightmare—unless you had an algorithm with the power to solve NP-complete problems.

No one would actually want to do such a thing. There is more to justice than a logically consistent legal system. But consider how breaking through this computational barrier would change the nature of mathematics. Showing that NP-complete problems can be solved in polynomial time is referred to in the trade as proving that NP=P—as powerful an equation as $E=mc^2$. Proving something like Fermat's last theorem, which was finally accomplished in 1995 with 108 pages of equations, is comparable to starting with a huge pile of jigsaw puzzle pieces and seeing if there is a way to assemble them into a pleasing picture. For 350 years mathematicians blundered through the space of possibilities, trying the pieces in various combinations, then the combinations with other combinations, reaching one impasse after another. Many had come to believe that a proof was impossible. But once someone finally came up with one, it was verified in a few months. That is the hallmark of an NP-complete problem: Finding an answer can be nigh impossible, but checking one is a breeze.

If it turns out that, against most expectations, NP=P, then a mathematical renaissance would be at hand. Generating a reasonably sized proof of a theorem (if one exists) would become as easy as verifying one. Armed with the amazing new algorithm, a computer could slice through problems that have gone unsolved for centuries.

So many creative endeavors involve exploring an exponentially expanding array of possibilities: assembling snippets of sound and silence into a musical composition, colors into a painting, words into a poem. Given one of these creations, it is relatively easy for an expert to verify that it is probably authentic, that it satisfies a list of criteria describing a certain artist's style. A writer favors particular word combinations and locutions, as well as plot elements and themes; an artist is partial to certain colors, subjects, and painting techniques; a composer to characteristic rhythms, melodic phrases, and keys. One can imagine that it might be possible, particularly with music, to formalize these rules, producing a program that would decide with a certain probability that a work is Bach-like—or maybe even Shakespearean or Van Gogh–esque. If NP=P, then perhaps creation would become as easy as verification: it would be equally possible to run the machine backward, to make new compositions that were unmistakably like what the artist would do.

. . .

Whether a quantum computer would be able to crack the domain of NP-complete remains one of the great open questions of science. The success of Peter Shor's factoring algorithm seems at first to offer some encouragement. Here, too, mathematicians were faced with a problem where each incremental increase in the number of variables caused the solution time to expand exponentially. With a quantum computer, Shor demonstrated, the task would be tamed, scaling instead in polynomial time.

Difficult as it is, factoring is believed to fall outside the class of NP-complete problems, though no one has proved this. There is still the possibility of the thirteen-year-old hacker or her father proving that factoring also belongs to this class. Then, voilà, the rest of the problems would fall as well: the traveling salesman problem, protein folding, software verification—providing, of course, that one could build a large-scale quantum computer.

Of course it is also still possible that something quite different will happen—that someone will discover a fast, efficient means of factoring large numbers on a classical computer. Then Shor's algorithm would be reduced to a curiosity. Simultaneously performing calculations in quantum superposition would still be a neat scientific trick, and an important tool in simulating particle physics. But for breaking a number into its primes, the technique would not be fundamentally more powerful than yoking together a bunch of Pentium chips.

Most theorists are pretty sure that won't happen. They are much less certain about quantum computing and the NP-complete class. Intuitively it seems that solving the traveling salesman problem would just be a matter of representing each of the possible routes as a row of 1s and 0s. For 15 cities there are 650 billion routes. That's about 2^{36} power, so all of these paths could be represented simultaneously with a row of 36 atoms pointing both up and down.

Apparently it is not going to be that easy. A paper published in 1997 by Umesh Vazirani, Ethan Bernstein, and Bennett and Brassard strongly suggests that quantum computing won't provide the exponential speedup needed to breach the realm of NP-complete. But the paper makes certain assumptions that may or may not turn out to be true. Shor, after all, didn't take the most obvious route to quantum factoring. Success came when he exploited a subtle means of mapping the problem onto another very different one involving, of all things, wave analysis. Who would have guessed that factoring a number and refracting a wave of numbers would be equivalent mathematical tasks? Hence some researchers remain optimistic that there may be a hidden structure deep within NP-complete problems—one that would allow their solution by quantum-mechanical means. No one knows. This is still unexplored terrain.

So how then do the proteins do it? Remember that even though a problem is intractable in general, there may be specific cases that are easily solved: The cities the salesman wants to visit might all lie on a straight line. Perhaps over the millennia, evolution has sifted through the universe of possible amino acid chains and favored those that can fold up easily.

That is the scientifically conservative explanation. At the other extreme is a suggestion made by a few theorists that proteins really are solving NP-complete problems every time they fold. And that would mean one of two things: Perhaps, against most expectations, the problem is not so hard after all. Once we find the algorithm the proteins use, we can program ordinary computers to do the same. Or perhaps the proteins are somehow exploiting some kind of exotic computation that people don't yet know about.

It is difficult to imagine that a protein could somehow harness quantum computation, tapping into the calculational power of its atoms. But very smart scientists have entertained ideas just as strange. The British physicist Roger Penrose believes that the human brain is itself a quantum computer, and that this explains the mystery of human consciousness—how three pounds of cerebral mush gives rise to the ineffable feeling of being alive and aware. Who knows? Maybe Oliver Sacks's mathematically precocious twins were unknowingly using quantum computation to pick prime numbers from the air.

There is so much we do not understand. With a mere one thousand atoms, a speck far too tiny to see, one could represent every number that is a thousand bits long, a thousand 1s and 0s. Translate that into decimal notation. Two to the 1,000th power is about 10^{301}. So every number from 0 to 9,999,999,999,999,999,999,999,999,999,999,999,999,999,999, 999,999,999,999,999,999,999,999,999,999,999,999,999,999,999, 999,999,999,999,999,999,999,999,999,999,999,999,999,999,999, 999,999,999,999,999,999,999,999,999,999,999,999,999,999,999, 999,999,999,999,999,999,999,999,999,999,999,999,999,999,999, 999,999,999,999,999,999,999,999,999,999,999,999,999,999,999, 999,999,999,999,999,999,999,999,999,999,999 could be held simultaneously in quantum superposition. Now run an algorithm that performs some computation. The result would be 10^{301} calculations all being carried out at the same time. But consider: That is vastly more strings of data than there are particles in the universe. So where is all that computing being done?

David Deutsch, one of the pioneering theorists in this field, believes the answer is clear: Each of these parallel calculations must be taking place in a separate universe. Thus even a modestly sized quantum computer with a few hundred qubits would stand as proof that the "many-worlds" interpretation of quantum mechanics is true. In one universe the photon in the two-slit experiment goes through the left-hand hole. In another universe it goes through the right-hand hole. Run Shor's algorithm to factor the number 1,000,000 and each of the different divisors is being processed in a parallel world.

Hardly anyone else finds this very convincing. The "many-worlds" interpretation is a popular way to think about quantum mechanics, a way to get one's mind around a phenomenon that no one can claim to truly understand. But the question of whether these hypothetical realms, the lands of might-have-been, are real places leads quickly into a shadowland where physics and philosophy collide.

Where do you draw the line between reality and its representation, between the map and the territory? That may be the hardest problem in the universe. Is it sensible to talk about proteins solving problems and carrying out algorithms? Or is that just a handy way to describe what is going on? Are the planets calculating the equations of Newton and Kepler as they course around the sun? Or is our mathematics just a shorthand for approximating the behavior of the physical world?

When computing was done with finely crafted brass gears and meticulously marked slide rules, the answer seemed clear. People designed these macroscopic objects—these great gobs of molecules—and then, with extreme difficulty, coaxed them into carrying out computations. There was nothing natural about it. Likewise, in an electronic computer, the streams of electrons are channeled through carefully laid wires. Computation is a very artificial endeavor, imposed on nature from the top down.

With quantum computing comes a fundamental shift. The ability to spin up and down, to hold multiple states in superposition, to become entangled—all this is what atoms naturally do. Performing a quantum computation is a matter of jumping on the wagon and going along for the ride.

So maybe this kind of computing has always been going on. The particles that have been bouncing around since creation, flipping each other's spins, exchanging information, have been performing some kind of great cosmic calculation. The universe might itself be a quantum computer. But that is another book.

Epilogue:
The Nine Billion Names of God

As the snows began to melt from the Jemez Mountains in New Mexico, workers were completing the new building where Q, one of the world's fastest classical computers, would reside. The one-acre floor began filling with rows of cabinets and miles of wires. Soon the cooling stacks were spewing the first heat of computation into the clear, cold sky.

Nearby in a small brown stucco building, physicists were still endeavoring to raise the threshold of quantum computation a few notches, from seven atoms to ten. And all over the world, scientists were hoping for the technological breakthroughs that might cause the expansion to take off exponentially—a new Moore's Law that would describe an arc leading in their lifetime to a working quantum computer. Then would come the excitement of seeing just what it could really do: factoring huge numbers, searching through data in record time, maybe (a long shot) even cracking the domain of NP-complete.

It may turn out that it is just not possible to scale up the tabletop experiments to anything nearly so grand. The late physicist Rolf Landauer at IBM used to delight in firing off vituperative e-mails to colleagues (and science writers) questioning the grander dreams for the infant technology. "When science journalists are celebrating the latest and most speculative proposal," he once admonished me, "I often urge them to do a 'Whatever happened to . . . ' item on some fad that has come and gone." His favorite example was an all-optical computer, in which information would be rapidly processed as pulses of light. There have been many ideas that looked promising on paper—fusion energy, high-temperature superconductivity, artificial intelligence—but whose real-world applications have been, so far, disappointing.

For many researchers, the most satisfying thing about quantum computing is the visceral feel it gives them for happenings in the subatomic realm. When you are trying to manipulate a few qubits, concepts like superposition and entanglement become much less abstract. Hands-on experiments like the ones at Los Alamos provide a tool to test quantum mechanics the way swinging pendulums and ricocheting billiard balls are used in high schools to test classical Newtonian mechanics. "Rather than teaching us how to build a large quantum computer," wrote two Parisian physicists, "such experiments are more likely to teach us about the processes that would ultimately make the undertaking fail." They meant that in a positive way. What could emerge is a deeper understanding not just of subatomic physics but of computation and its role in the basic workings of the natural world. That may be more valuable than salable machines.

If the field does take off and the domain of NP-complete falls to quantum mechanics, other challenges would remain—fuel perhaps for even more exotic theories of computing. There are signs that the games of chess, checkers, and go are in an even harder class than NP-complete. If so, then searching through all the possible plays and counterplays for a winning strategy would not be possible in anything faster than exponential time. It has been proven that analyzing one of these games for an arbitrarily large board—the kind of generalization that mathematicians love—is indeed exponentially hard. (It's fun to imagine a contest like this, white and black chasing each other across an endless checkered expanse.) It remains an open question whether real, finite versions of the games are also intractable.

Also in the category of harder-than-NP-complete is the ability to prove any mathematical theorem, no matter what its length. It has been established that this is impossible—that there is no conceivable algorithm—if the statements are expressed in a powerful mathematical language called predicate calculus. ("All C are B and no B are A, so no C are A.") If some restrictions are placed on the kinds of expressions the computer is asked to handle, algorithms may exist. But they scale at a nightmarish rate. As the theorem to prove grows in length and complexity, with more and more variables to juggle, the solution time increases not just exponentially but *superexponentially.* That means that the difficulty rises at the rate of something like 2 to the power of 2 to the power of n, the exponents stacked up like tiers on a wedding cake. Here n is a measure of the length of the problem. Increase it notch by notch from 1 to 9 and the solution time jumps from 4 (2 to the power of 2^1) to 16 (2 to the power of 2^2) to 256 (2 to the power of 2^3) . . . to 65,536 to 4,294,967,296 and then 1.8×10^{19}, 3.4×10^{38}, 1.2×10^{77}, and 1.3×10^{154}. In less than a dozen steps, one reaches a point where a computer the size of the universe would take longer to solve the problem than the universe is expected to exist.

Finally there are problems that would require an infinite amount of time. This, mathematicians suggest, is as close to a definition of impossible as one could ask for. Remember the Turing machine, the contraption with the paper tape that is mathematically equivalent to any possible digital computer? Given any solvable problem, the machine will eventually grind out a solution. If there is none, it will churn and churn forever. The dilemma is knowing when to give up. Just as you are ready to pull the plug after a few hundred years, the machine might be on the verge of an answer.

What if one could devise an algorithm that could examine any Turing machine and its tape and determine whether it would eventually halt? Then at least the unsolvable problems could be weeded out at the beginning. Alas, Turing himself proved that to be impossible: the program for solving the halting problem would itself not halt. Even the exponential speedup afforded by quantum computation would be no help. Infinity is still infinity no matter how fast it flies by.

Again, there may conceivably be solutions, perhaps requiring some unimaginable kind of computation involving super-strings—the physicists' latest candidate for the tiniest constituents of matter. But here we edge into science fiction.

From Los Alamos it is a 36-mile drive up and across the Valle Grande to Jemez Springs, home of both a Catholic retreat and a Zen Buddhist monastery. The abrupt change of locale and spiritual ambience brings to mind a science fiction story published in the 1950s by Arthur C. Clarke.

It begins when a Tibetan lama comes to New York City to purchase the latest in computing machines, called the Mark V. He explains to a rather bewildered company executive that the monks of his Himalayan lamasery believe that mankind's earthly mission is to list all the names of God. Thus for three centuries their scribes have been writing down every possible permutation of letters in a special alphabet they have devised. (For some reason they believe that all the many names can be written with a string of no more than nine letters.)

They had expected the task to take about 15,000 years—until they learned about digital computers. With the Mark V they could produce every possible combination in a matter of months. Most of them would be junk, like the output of the monkeys with the typewriters. But somewhere in the pages they would know they had spelled out every one of the names of God.

The company agrees to ship the machine (the lama has plenty of money in an Asian account) along with two engineers to supervise.

Cut now to the Tibetan highlands, three months later, where Project Shangri-la, as the bemused Westerners call it, is nearing completion. Powered by a diesel generator (also used to spin the monastery's prayer wheels) the Mark V has been spewing forth page after page of symbols, which the monks dutifully paste in large ledger books.

The climax to the story nears as one of the engineers, named Chuck, learns a disturbing truth: Once the last of the names has been written, the monks believe, the world will come to an end. The purpose of existence will have been accomplished. At first this seems funny, but slowly he and his partner, George, begin to worry. How will the monks react when the last page comes off the printer and nothing cataclysmic happens? Will they become violent and blame the computer's keepers? They decide not to stay around to find out.

In the final scene they are riding down the winding trail on their mountain ponies, heading toward the valley and a plane back to the United States. "Wonder if the computer's finished its run?" George says. "It was due about now."

Chuck didn't reply, so George swung round in his saddle. He could just see Chuck's face, a white oval turned toward the sky.

"Look," whispered Chuck, and George lifted his eyes to heaven. (There is always a last time for everything.) Overhead, without any fuss, the stars were going out.

The end.

For the monks, getting a 1950s-era digital computer allowed them to suddenly do the unthinkable. Our civilization has its own impossible dreams. Today Clarke could rewrite the story, replacing the monks with computer scientists obsessed with intractably hard problems. And standing in for the Mark V would be a quantum computer. Just what this could mean for science will become clearer, by and by. For now the question remains in the speculative realm where writers like Clarke excel. Even the best scientists don't know how this story will turn out. You don't have to be a computer geek to find something rather exhilarating about that.

The Fine Print: Notes and Sources

Writing this kind of book is like having two very different people breathing down your neck. The scientist keeps trying to grab away the keyboard to qualify every other sentence with footnotes and equations. Meanwhile, the reader is prodding you to cut to the chase and get on with the story. These notes, I hope, are a way of accommodating them both while giving me a chance to talk back.

I will also use these pages to suggest further reading and to cite some of the important papers in the field. Additional bibliographic references can be found in the more technical books about quantum computing mentioned at the end of the notes section.

Preface: Inside the Black Box

xv   Alan Lightman's remarks appeared in a nice essay of his own introducing *The Best American Essays 2000* (Houghton Mifflin).

Prologue: The Road to Blue Mountain

5   the new Los Alamos supercomputer: When up and running, Q will already have been surpassed, just barely, by a 35.6-teraops computer brought online in 2002 by the Japanese at the Earth Simulator Research and Development Center in Yokohama. And, not to be outdone, Los Alamos's sibling laboratory, Lawrence Livermore in California, is building its own new supercomputing center, which in coming years is to hold two 100-teraops machines.

6   a quantum switch can paradoxically be in both states at the same time: People who frequently read popular accounts of physics have probably noticed that certain adjectives, like "paradoxical" and "schizophrenic," are almost obligatory for describing quantum behavior. Some physicists react to this kind of language as they would to the sound of a squeaking balloon. They prefer to speak of things like state vectors, eigenvalues,

and other tools of the trade. It is a little embarrassing to keep pulling out the same rhetorical stops, but it serves a purpose: to avoid sapping too much brain power straining for the meaning of philosophical implications nobody can really understand.

7   the Los Alamos experiment: More precisely, it harnesses the *nuclei,* or cores, of the atoms to compute. (The orbiting electrons are, in this sense, irrelevant.) But this early on, it doesn't hurt to just say that it's the atoms that are being used as the "quantum abacus beads." More details are in a later chapter.

8   all things being equal a supercomputer would occupy 5,000 Earths: Those are important weasel words. With Q we have about 12,000 processors in a space of half an acre. So say that the full one-acre floor would hold 24,000 processors, and roughly speaking, the whole computer would do that many calculations at the same time. So to do 18 quintillion calculations the area would expand by a factor of 18×10^{18} divided by 24×10^3, which comes out to about 750 trillion acres. A square mile is 640 acres, so we end up with more than a trillion square miles, 5,000 times the size of the surface of the Earth. Now actually, a single processor (though basically a serial calculator) can perform more than one operation during each machine cycle, so maybe the imaginary machine would occupy merely a thousand Earths. And perhaps before long the processors will be ten times faster. So that brings us down to a hundred Earths. That's how it goes with these back-of-the-envelope calculations. The point of all the arithmetic is just to say that it would be very big indeed.

9   the difficulty of factoring large numbers: Around the time of this writing, people were commonly estimating that finding the factors of a 400-digit number would take billions of years. But it's risky to be so specific. In the late 1970s, a very smart mathematician could say, with little fear of contradiction, that factoring a 125-digit number would take 40 quadrillion years. Then in 1994 someone managed to factor a number 129 digits long. (This is from page 129 of *Applied Cryptography* by Bruce Schneier [Wiley, 1996].) It is easy to underestimate the speedup in computing power or the cleverness of computer scientists in finding calculational shortcuts. But none of this changes the fact that each time you expand the number by a single digit, the factoring time grows exponentially (or, more precisely, "superpolynomially," which is still very bad—more on this later). Finding a way around that inconvenient fact would merit a Nobel Prize in mathematics, if there were such a thing.

1. "Simple Electric Brain Machines and How to Make Them"

12   cipher and decipher codes: I've corrected a mistake in the ad, which said
     cipher and *encipher*, which of course mean the same thing. At the time
     of this writing, Geniac ads could be found at http://psych.butler.edu/
     bwoodruf/pers/geniac/geniacadvertisement.htm and the manual and
     other   documents   at   http://www.computercollector.com/archive/
     geniac/.

15   "How would you put a thread in a small hole?": Choice (b) assumes one
     knows that a tap is the name of a tool for making threaded screw holes.

18   the Net is just a bunch of Geniacs: Or, as Alan Turing showed in his
     famous 1937 paper, "On Computable Numbers, with an Application to
     the *Entscheidungsproblem*," the computer is a universal machine.

2. Tinkertoy Logic

19   cities compared to computer chips: Thomas Pynchon may have been
     the first to turn the metaphorical tables in his novel *The Crying of Lot 49*
     (Lippincott, 1966).

20   the Tinkertoy computer: For more details, see Daniel Hillis's *The Pat-
     tern on the Stone* (Basic Books, 1998), which is also an excellent intro-
     duction to the powerful concepts of computer science, and A. K.
     Dewdney's "Computer Recreations" column in the October 1989 issue
     of *Scientific American* (reprinted in his collection, *The Tinkertoy Com-
     puter and Other Machinations*, Freeman, 1993). Though out of print,
     Jeremy Bernstein's 1964 book-length essay, *The Analytical Engine* (Mor-
     row, reprinted 1981), is still a fine short history of the computer.

3. Playing with Mirrors

32   Moore's Law, which might be more accurately called Moore's Observa-
     tion or Moore's Prediction—there is nothing ironclad about it—was
     stated by Intel cofounder Gordon E. Moore in 1965 when he was direc-
     tor of research and development at Fairchild Semiconductor. As he put
     it, the number of transistors and other parts that could be squeezed
     onto an integrated circuit chip would continue doubling annually for at
     least the next decade. To get it straight from the horse's mouth, see his
     article "Cramming More Components onto Integrated Circuits" (*Elec-
     tronics* 38, no. 8 [April 19, 1965]). One reason the law has continued to

hold is that it has been conveniently amended to say that this density doubles about every eighteen months. As long as the increase is exponential, it remains significant. Consider: If you could fold a piece of typing paper in half a hundred times, it would be many light-years thick.

34   Φ: This symbol, solely because it looks like a 1 superimposed on a 0, will be used in the book to mean any combination of those two states. For, as noted later on, a superposition can be, say, 75 percent 1 and 25 percent 0, or 76 and 24, or 77 and 23, and so on. There are any number of possible blends. This results in all kinds of subtleties that will be glossed over here to emphasize the key idea (which is difficult enough): that a quantum object can be in two opposing states at the same time. I see that Φs also pop up in the formal equations of quantum mechanics, but as far as I can tell, I am using the symbol in a different way.

34   Feynman, Benioff, and early hints of quantum computation: Feynman laid out his views in "Simulating Physics with Computers" (*International Journal of Theoretical Physics* 21, no. 6/7 [1982], pp. 467–88), and "Quantum Mechanical Computers" (*Foundations of Physics* 16 [1986], pp. 507–31); Benioff in "Quantum Mechanical Hamiltonian Models of Turing Machines" (*Journal of Statistical Physics* 29 [1982], pp. 515–46). Just a few years later David Deutsch published another seminal paper: "Quantum Theory, the Church-Turing Principle, and the Universal Quantum Computer" (*Proceedings of the Royal Society of London*, A400 [1985], pp. 96–117).

35   a hot, glowing object would radiate infinite light: This was known as the "ultraviolet catastrophe."

35   the beginnings of quantum theory: Planck basically showed that light is absorbed in packets, Einstein that it is emitted that way.

36   the photoelectric effect: More specifically, Einstein showed that the speed of these particles depended not, as might be expected, on the brightness of the light but on its frequency: how fast it vibrated.

36   five or six photons to fire the human retina: This refers to dim, monochromatic light and is taken from Feynman's book *QED: The Strange Theory of Light and Matter* (Princeton University Press, 1985). This has become my absolute favorite introduction to quantum mechanics. Two other first-rate popular treatments are David Lindley's *Where Does the Weirdness Go?: Why Quantum Mechanics Is Strange, but Not as Strange as You Think* (Basic Books, 1996) and Nick Herbert's *Quantum Reality: Beyond the New Physics* (Anchor/Doubleday, 1985). Brian Silver has a

clear and nicely compact forty-three-page overview of the field in his history *The Ascent of Science* (Oxford University Press, 1998), chapters 28 through 30. Sam Treiman's *The Odd Quantum* (Princeton University Press, 1999) is highly regarded among introductory books on the next rung up, but it's quite a leap higher.

36   photons bouncing from windowpanes: Feynman uses this example in QED, pointing out that the phenomenon can also be explained by thinking of light as an old-fashioned wave.

38   wavelets each describing one of the particle's many possible states: Each gives a number, called the "amplitude," which is the square root of the probability. That detail will become important later on, but for now just think of a wavelet as representing a possibility.

38   the two-slit experiment: This has become one of the tired old work-horses of science writing. But since I haven't come across anything that works better, I reluctantly trot it out.

44   experiments confirming the EPR effect: The first, in 1982, was carried out by Alain Aspect in Paris. Like Einstein, many people have tried to explain away such results by embellishing quantum theory with "hidden variables"—unknown factors that would account for the particles' seemingly mysterious connection. But ultimately these all rely on faster-than-light, or "superluminal," signaling. That is the tradeoff. Any way you look at it, quantum mechanics is very weird.

48   one quantum system simulating another: What Feynman was describing can be thought of as the quantum version of an analog computer.

4. A Shortcut Through Time

53   Turing machines: Andrew Hodges, author of the celebrated biography *Alan Turing: The Enigma* (Walker, 2000 reissue), maintains a wonderful Web site on the subject: http://www.turing.org.uk/turing. It includes pointers to other sites where you can play with virtual Turing machines.

55   No device *operating according to classical physics* can do better than a Turing machine: This is essentially what is known as the Church-Turing hypothesis (after Alonzo Church and Alan Turing): Anything that can be reasonably considered to be computable can be computed on a Turing machine.

56   mathematical savants: "The Twins," in Oliver Sacks, *The Man Who Mistook His Wife for a Hat* (Touchstone, 1998 reissue), pp. 195–213.

57   Driving 100 miles at a constant velocity takes twice as long as 50 for a
      BMW or a VW Bug: Of course if acceleration is allowed then the BMW
      will win. The example would then no longer be linear.

58   polynomial functions: Strictly speaking the term means "many-named"
      and refers to any equation whose terms are raised to integral nonnega-
      tive powers—x^2, z^4, and so forth. These are the problems mathemati-
      cians call "tractable." Functions that scale faster (i.e., are more
      "computationally complex") are collectively called "superpolynomial";
      this includes but is not limited to the exponential functions. Confus-
      ingly, "superpolynomial" is also used to refer to functions that are in
      between polynomial and exponential in complexity, and "exponential"
      is regularly used to mean anything faster than polynomial, i.e., anything
      that is *intractable*.

60   the factoring prizes: One hundred fifty-five decimal digits is 512 digits
      in binary, and 617 decimal digits is 2,048 in binary. For more infor-
      mation see the site for the RSA Factoring Challenge: http://www.
      rsasecurity.com/rsalabs/challenges/factoring/faq.html.

63   a quantum Turing machine: The idea was described by David Deutsch
      in 1985 in "Quantum Theory, the Church-Turing Principle, and the
      Universal Quantum Computer" (see the notes for chapter 3 for the full
      citation). This might be considered the date that the quantum comput-
      ing field began in earnest.

65   quantum mechanics as a shortcut through time: To think of it another
      way, a regular old Turing machine could simulate a quantum computer
      (do anything it could do) but only in exponentially expanding time.
      That, of course, is the whole point. It is important, my computer science
      consultant notes, to emphasize the distinction between computational
      complexity—how fast a problem scales—and *computability*, whether a
      problem is solvable at all (the Turing machine might gyrate and flail
      forever without converging on an answer). Quantum computers out-
      strip classical computers in computational complexity, but as far as is
      known they can't solve problems that are uncomputable in any amount
      of time by a Turing machine.

5. Shor's Algorithm

67   the classical cellular automaton (CA) as a headless Turing machine: In
      his monumental work, *A New Kind of Science* (Wolfram Media, 2002),
      Stephen Wolfram lays out a proof that a simple one-dimensional, two-

state CA (i.e., one like the ones in this book, with a single row of cells that can be either black or white) is capable of emulating a universal Turing machine. This brings new weight to the increasingly popular notion that the universe itself is computational. I write more about this in *Fire in the Mind* (Alfred A. Knopf, 1995).

68   CAs like these pictured in this chapter can be generated with a neat simulator found on a Web site set up by Andreas Ehrencronas: http://cgi. student.nada.kth.se/cgi-bin/d95-aeh/get/life?lang=en.

68   the tension between order and chaos: I'm using "chaos" here in the old-fashioned dictionary sense meaning "a state of things in which chance is supreme" (*Merriam-Webster's Collegiate Dictionary,* 10th ed.), rather than limiting it to the definition used in the study of nonlinear mathematics, i.e., systems that are extremely sensitive to tiny changes in their initial conditions (the so-called "butterfly effect").

71   the quantum cellular automaton: I was first struck by this notion in the early 1990s after hearing a lecture at the Santa Fe Institute by Seth Lloyd, who is now a professor at the Massachusetts Institute of Technology. See his paper, "A Potentially Realizable Quantum Computer" (*Science* 261 [September 17, 1993], pp. 1569–71).

73   Shor's algorithm is described most completely in his paper "Polynomial-time Algorithms for Prime Factorization and Discrete Logarithms on a Quantum Computer" (*SIAM Journal on Computing* 26, no. 5 [1997], pp. 1484–1509. SIAM stands for Society for Industrial and Applied Mathematics). This is an updated version of a 1994 paper in which he first described the factoring method. I've tried to lay out the steps of this complicated procedure as clearly and economically as possible, but there is a point—where the quantum computer performs the Fourier transform—at which I had to engage in a little handwaving. (I'm reminded of an old cartoon by Sidney Harris, which is available on a T-shirt from the American Institute of Physics: Two scientists are gathered around a blackboard filled with esoteric equations and the words THEN A MIRACLE OCCURS. One says to the other, "I think you should be more explicit here in step two.")

79   Shor knew of research on quantum Fourier methods: Specifically a paper by Dan Simon of the University of Montreal: "On the Power of Quantum Computation" (*SIAM Journal on Computing* 26, no. 5 [1997], pp. 1474–83). Ethan Bernstein and Umesh Vazirani at the University of California at Berkeley had also come upon the idea.

6. Breaking the Code

84   "Every letter in the English language": This appears on page 23 of Simon Singh's *The Code Book: The Science of Secrecy from Ancient Egypt to Quantum Cryptography* (Anchor Books, 1999), a clear and entertaining account of the subject. Another good one is *The Codebreakers* by David Kahn (Scribner, revised edition, 1996). Written in a more technical (but amusingly iconoclastic) style is Bruce Schneier's *Applied Cryptography* (Wiley, 1996).

88   public key cryptography: I have skipped over an important precursor to RSA, a system developed by Whitfield Diffie, Martin Hellman, and Ralph Merkle and based on modular or "clock" arithmetic. Actually, in addition to prime numbers, RSA also employs clock arithmetic, a detail tangential to our purposes here. Singh explains it in an appendix to *The Code Book*.

90   Lloyd's demonstration that Feynman's idea was sound: "Universal Quantum Simulators" (*Science* 273, no. 5278 [August 23, 1996], pp. 1073–78).

90   "pseudorandom" number generators: My favorite is driven by the digitized output from lava lamps. See "Connoisseurs of Chaos Offer a Valuable Product: Randomness," by George Johnson, *The New York Times*, June 12, 2001.

91   a "plug and play" Quantum Random Number Generator: The company is called id Quantique. Its products, including a device for quantum cryptography (wait for chapter 9), are described at http://www.idquantique.com.

92   Grover's quantum search method is described in several papers, including "Quantum Mechanics Helps in Searching for a Needle in a Haystack" (*Physical Review Letters* 79, no. 2 [July 14, 1997], pp. 325–28) and "A Fast Quantum Mechanical Algorithm for Database Search" (*Proceedings of 28th Annual ACM Symposium on Theory of Computing* [May 1996], pp. 212–19).

93   Shannon's calculation on the intractability of chess: "A Chess-Playing Machine" (*Scientific American,* February 1950, pp. 48–51).

95   an event with a probability of 9: Actually probability is usually expressed on a scale of 0 to 1. So even odds (50-50) would be .5 (and the amplitude would be the square root of that: 2.24 or −2.24). A probability of 100 percent is expressed as 1.

101   the basic processing could be done with a row of 20 qubits: There would
      actually have to be several times that many to account for error correc-
      tion. I get into this in chapter 8.

103   "writing the software for a device that does not yet exist": This is from a
      very clear popular article by Grover in the July-August 1999 issue of *The
      Sciences,* a prizewinning magazine that the New York Academy of Sci-
      ences stopped publishing in 2001.

104   the Eniac on a chip: http://www.ee.upenn.edu/~jan/eniacproj.html.

7. Invisible Machines

105   making an AND gate from an OR and three NOTs: You can do this at
      home. Put two of the NOT gates on each of the inputs of the OR gate
      and the third one on the output.

106   information is always conserved in the quantum realm: Just how it is
      that completely reversible reactions in the subatomic substrate give rise
      to a macroscopic world where many things cannot be undone is one of
      the real puzzles of physics. This is another subject I get into more deeply
      in *Fire in the Mind* (cited earlier), along with the whole notion that
      information may be as fundamental as matter and energy.

107   Landauer's Principle: R. Landauer, "Irreversibility and Heat Generation
      in the Computing Process" (*IBM Journal of Research and Development* 5
      [1961], pp. 183–91).

108   Fredkin gates and Toffoli gates: These are named for Edward Fredkin
      and Tommaso Toffoli, two of the best-known theorists in the informa-
      tion-is-physical movement. For an elegantly funny account of all this,
      look for Robert Wright's classic book, *Three Scientists and Their Gods:
      Looking for Meaning in an Age of Information* (Times Books, 1986)—a
      valuable work that has been allowed to go out of print.

112   spelling out "IBM" with atoms: It looks something like this:

```
  • • •    • • •       •      •
  •      •    •     • •  • •
  •        • • •      •   •  •
  •      •     •     •       •
  • • •    • • •      •      •
```

      You can see images of this and other "atomist" art (like "Carbon
      Monoxide Man," made from CO atoms arranged on a platinum canvas,
      and the "Quantum Corral," iron on copper) at IBM's STM Gallery:
      http://www.almaden.ibm.com/vis/stm/gallery.html. The initials mean

"scanning tunneling microscope," the instrument used for the manipulation.

113   Cirac and Zoller's quantum gate: "Quantum Computations with Cold Trapped Ions" (*Physical Review Letters* 74, no. 20 [May 15, 1995], pp. 4091–94).

114   rocking atoms: More specifically, 0 and 1 represent whether the atoms are rocking in their lower- or higher-energy mode.

116   the NIST lab: http://www.bldrdoc.gov/timefreq/ion/index.htm.

117   the quantum cat: Stephen Hawking famously said, "When I hear about Schrödinger's Cat, I reach for my gun." I feel the same way at times, yet here I am recycling the story again. Note that, as with many of these old tales, there are various versions involving different mechanisms, but the lesson is the same: There is a mismatch between how things work on the quantum and the classical realms.

118   decoherence: David Lindley explains this nicely in *Where Does the Weirdness Go?* cited earlier.

118   NIST's cat experiment: C. Monroe, D. M. Meekhof, B. E. King, and D. J. Wineland, "A Schrödinger Cat Superposition State of an Atom" (*Science* 272 [May 24, 1996], pp. 1131–36). Note that this was actually published after the ion-trap experiment. I mention it first for didactic reasons.

118   the ion-trap switch: C. Monroe, D. M. Meekhof, B. E. King, W. M. Itano, and D. J. Wineland, "Demonstration of a Fundamental Quantum Logic Gate" (*Physical Review Letters* 75, no. 25 [December 18, 1995], pp. 4714–17).

120   By tuning the beam just so: This is a very useful locution when you want to avoid mucking around in details that would only distract from the thrust of the story. If you want to know the specifics, see the paper cited in the previous note.

120   "multiplexing" ion traps: A scheme (a little different from the one suggested here) is described in D. Kielpinski, C. R. Monroe, and D. J. Wineland, "Architecture for a Large-Scale Ion-Trap Quantum Computer" (*Nature* 417, no. 6890 [June 13, 2002], pp. 709–11).

8. Counting with Atoms

122   Cavity QED: The labs at Caltech and the École Normale Supérieure are run, respectively, by Jeffrey Kimble (http://www.its.caltech.edu/~qoptics) and Serge Haroche (http://www.lkb.ens.fr/recherche/qedcav/

english/englishframes.html). Here are two seminal papers on this work: Q. A. Turchette, C. J. Hood, W. Lange, H. Mabuchi, and H. J. Kimble, "Measurement of Conditional Phase Shifts for Quantum Logic" (*Physical Review Letters* 75, no. 25 [December 18, 1995], pp. 4710–13; this was published adjacent to the previously mentioned Monroe and Wineland paper); and P. Domokos, J. M. Raimond, M. Brune, and S. Haroche, "Simple Cavity-QED Two Bit Quantum Logic Gate: The Principle and Expected Performances" (*Physical Review A* 52 [November 1995], pp. 3554–59).

122   "flying qubits": It is Jeff Kimble at the California Institute of Technology who coined this term. (Artur Ekert at Oxford tells me he actually means it to refer to photons propagating through empty space rather than ones confined in the experimental cavities.)

123   it is easy to become bogged down with details: From the Haroche paper: "We consider here a Fabry-Pérot type cavity, with a Gaussian transverse beam profile (waist $w = 5.96$ mm). The experimental values of the field energy damping time t_{cav} for such cavities range from 1 to 30 ms. . . . To be more specific . . . the le>—lg> transition at frequency $\omega_0/2\pi = 51.099$ GHz is quasiresonant with the cavity mode at frequency $\omega/2\pi$."

124   NMR computing with carbon atoms: Actually, since carbon has twelve protons and neutrons—an even number—researchers use carbon-13, an isotope with an extra neutron, to get a net spin.

130   a quantum computer runs Grover's and Shor's algorithms: L. M. K. Vandersypen, M. Steffen, M. H. Sherwood, C. S. Yannoni, G. Breyta, and I. L. Chuang, "Implementation of a Three-Quantum-Bit Search Algorithm" (*Applied Physics Letters* 76, no. 5 [January 31, 2000], pp. 646–48) and, by the same authors, "Experimental Realization of Shor's Quantum Factoring Algorithm Using Nuclear Magnetic Resonance" (*Nature* 414, no. 6866 [December 20/27, 2001], pp. 883–87).

131   quantum dots: G. Burkard, D. Loss, and D. P. DiVincenzo, "Coupled Quantum Dots as Quantum Gates" (*Physical Review B* 59, no. 3 [January 15, 1999], pp. 270–74). For more on possible designs for solid-state quantum computers, see David DiVincenzo's "Real and Realistic Quantum Computers" (*Nature* 393, no. 6681 [May 14, 1998], pp. 113–14). Qubits, by the way, need not consist of single particles. Under the right conditions rings of superconducting currents can be put into superposition—flowing clockwise and counterclockwise. Seth Lloyd told me that a controlled NOT gate realized this way is "right around the corner."

132   A scientist at Bell Labs has proposed a device: His name is Phil Platzman. See P. M. Platzman and M. I. Dykman, "Quantum Computing with Electrons Floating on Liquid Helium" (*Science* 284, no. 5422 [June 18, 1999], pp. 1967–69).

135   quantum error correction: A. Berthiaume, D. Deutsch, and R. Jozsa, "The Stabilisation of Quantum Computations," in *Proceedings of the Workshop on Physics and Computation,* PhysComp 94 (IEEE Computer Society, Los Alamitos, Calif., 1994), pp. 60–62.

137   Steane's and Shor's advances: The results came in two independent papers: P. W. Shor, "Scheme for Reducing Decoherence in Quantum Computer Memory," *Physical Review A* 52, no. 4 (October 1995), pp. 2493–96, and Andrew Steane, "Error Correcting Codes in Quantum Theory," *Physical Review Letters* 77, no. 5 (1996), pp. 793–97.

137   "is essentially a way of embedding": see http://www.theory.caltech.edu/~quic/errors.html.

140   Los Alamos quantum error-correction experiment: E. Knill, R. Laflamme, R. Martinez, and C. Negrevergne, "Benchmarking Quantum Computers: The Five-Qubit Error Correcting Code" (*Physical Review Letters* 86, no. 25 [June 18, 2001] pp. 5811–14).

9. Quantum Secrecy

141   an unbreakable classical code: This is known in the trade as a one-time pad. You use the key once and throw it away.

142   quantum anticounterfeiting scheme: Stephen Wiesner's circa 1970 paper on quantum banknotes was belatedly published as "Conjugate Coding" (*SIGACT News* 15, no. 1 [1983], pp. 78–88).

146   Bennett and Brassard's original paper: "Quantum Cryptography: Public Key Distribution and Coin Tossing," in *Proceedings of IEEE International Conference on Computers Systems and Signal Processing* (Bangalore, India, December 1984), pp. 175–79.

149   Bob-and-Alice error detection: This can be done more efficiently with a parity-checking scheme. See C. H. Bennett, G. Brassard, and A. K. Ekert, "Quantum Cryptography" (*Scientific American* [October 1992], pp. 50–57).

151   id Quantique: This is the same company that makes the quantum random number generator described in chapter 6.

151   exchanging a quantum key through open air: W. T. Buttler, R. J. Hughes, S. K. Lamoreaux, G. L. Morgan, J. E. Nordholt, and C. G. Peterson, "Daylight Quantum Key Distribution Over 1.6 km" (*Physical Review Letters* 84, no. 24 [June 12, 2000], pp. 5652–55).

151   Eve skimming off photons: This is called "a beam splitter attack."

151   the fainter they are, the shorter the distance they can be sent: And conversely, if the flashes are too bright, consisting of many photons, each pulse will act like classical information, which can be copied without disturbing it. Gilles Brassard tells me that each dim flash, on average, contains perhaps one-tenth of a photon, an idea I find rather difficult to grasp. He explains that for every ten pulses, nine can be empty with one consisting of a single photon, and perhaps one flash in a hundred containing more than one photon.

152   Ekert's EPR scheme: "Quantum Cryptography Based on Bell's Theorem" (*Physical Review Letters* 67, no. 6 [August 5, 1991], pp. 661–63). Alice and Bob actually use three different orientations: 0 versus 90 degrees, 30 versus 120, and 60 versus 150.

153   Bell's Inequality (after the Irish physicist John Bell): We have seen the extreme case of stronger quantum correlations: If the two filters are held with the same orientation, separated by an angle of 0 degrees, Bob's 1s are invariably Alice's 0s. In other words, their measurements agree 0 percent of the time. Conversely, if Bob holds his filter horizontally and Alice holds hers vertically, their measurements will be identical, agreeing 100 percent of the time. But consider the in-betweens. If their filters are 30 degrees apart, their measurements will agree three times out of four—more often than possible in the classical world. Weird, but as Feynman put it, that's just the way it is.

153   quantum teleportation: C. H. Bennett, G. Brassard, C. Crepeau, R. Jozsa, A. Peres, and W. Wootters, "Teleporting an Unknown Quantum State via Dual Classical and Einstein-Podolsky-Rosen Channels" (*Physical Review Letters* 70, no. 13 [March 29, 1993], pp. 1895–99).

153   emitting photons one at a time: Z. Yuan, B. E. Kardynal, R. M. Stevenson, A. J. Shields, C. J. Lobo, K. Cooper, N. S. Beattie, D. A. Ritchie, and M. Pepper, "Electrically Driven Single-Photon Source" (*Science* 295, no. 5552 [January 4, 2002], pp. 102–5).

154   freezing light to a standstill: A. V. Turukhin, V. S. Sudarshanam, M. S. Shahriar, J. A. Musser, B. S. Ham, and P. R. Hemmer, "Observation

of Ultraslow and Stored Light Pulses in a Solid" (*Physical Review Letters* 88, article no. 023602 [January 14, 2002].

10. The Hardest Problem in the Universe

155   "nature's robots": This, in fact, is the title of a recent book on proteins by Charles Tanford and Jacqueline Reynolds (Oxford University Press, 2002).

157   Every two years they hold a competition: It is called CASP, for Critical Assessment of Techniques for Protein Structure Prediction, and takes place in alternate Decembers at the Asilomar Conference Center, outside Pacific Grove, California. More information is at http://prediction-center.llnl.gov.

157   the protein-folding problem opens up a world of philosophical debate: See, for example, Joseph Traub, "On Reality and Models," in *Boundaries and Barriers: On the Limits to Scientific Knowledge,* edited by John Casti and Anders Karlqvist (Addison-Wesley, 1996).

159   the definition of NP-complete: Here is the bigger picture: NP is the class of all problems, no matter how easy or how hard, whose solutions can be quickly verified—i.e., shown to be correct—in polynomial time. Within that class is P, problems that are easily solved, and NP-complete, problems that are difficult. As noted later in the text, NP-complete problems have another important characteristic: If any one of them can be solved in polynomial time, then so can they all.

159   If one of these electronic oracles could be built: In theorizing about what is computable and what is not, scientists often use the term "oracle" for a black box that, like the Oracle at Delphi, can step in at a crucial point of a computation and pull an answer from thin air. The implication is that if there were a way to replace the imaginary procedure with a real computer algorithm, the problem in question could be solved.

162   a software verification machine: For all its power to prove internal consistency, this imaginary device still wouldn't be able to guarantee perfect software. Gödel's Incompleteness Theorem proves that it is impossible to determine whether a logical system is both consistent and complete. You can have one or the other but not both. A program that was bugless would be incapable, by definition, of carrying out every single task intended by its designers. Conversely, if it fulfilled all its performance specifications, it would inevitably contain an internal flaw.

162   Fermat's last theorem: "There are no natural numbers x, y, and z such that $x^n + y^n = z^n$, in which n is a natural number greater than 2." It was proved in 1995 by Andrew John Wiles: "Modular Elliptic Curves and Fermat's Last Theorem" (*Annals of Mathematics* 141, no. 3 [1995], pp. 443–551). Actually the proof can be considered even longer, for it was accompanied by a crucial supplementary article written with Richard Taylor: "Ring-Theoretic Properties of Certain Hecke Algebras."

163   If NP=P, then creation would conceivably become as easy as verification: David Deutsch refers to this as "generating new Mozart" (see Julian Brown's *Minds, Machines, and the Multiverse* [Simon and Schuster, 2000], p. 296).

164   quantum computers and NP-complete: C. H. Bennett, E. Bernstein, G. Brassard, and U. Vazirani, "Strengths and Weaknesses of Quantum Computing" (*SIAM Journal on Computing* 26, no. 5 [1997], pp. 1510–23).

165   So how then do the proteins do it?: Another theory is that they are guided in the folding by helper molecules called "chaperones." In the paper cited earlier Traub lists other possibilities: the assumptions computational biologists make in their models might be wrong; our model of computation (the Turing machine) might be incorrect; it may not be necessary for the proteins to "solve" the problem exactly—a quick-and-dirty approximation may do. Traveling salesmen are able to get by without mathematically perfect itineraries.

166   Deutsch and the multiverse: He describes his worldview in his book *The Fabric of Reality: The Science of Parallel Universes—and Its Implications* (Penguin, 1998).

167   Hardly anyone else finds this very convincing: See a paper by Deutsch's Oxford colleague, Andrew Steane: "A Quantum Computer Only Needs One Universe," available on the Internet at the LANL Preprint Archive for Quantum Physics, http://arxiv.org/abs/quant-ph/0003084. There is a continuing debate over where the computational power of quantum mechanics comes from. Steane questions the common view: "Quantum superposition does not permit quantum computers to 'perform many computations simultaneously' except in a highly qualified and to some extent misleading sense." He and others argue that entanglement is more important than superposition. It's sobering, the deeper one gets into a subject, to learn how many fundamental ideas are still up in the air.

Epilogue: The Nine Billion Names of God

168 "When science journalists are celebrating the latest": This is from an e-mail Landauer sent me on January 8, 1997. He expressed these kinds of ideas in several papers, for example, "Is Quantum Mechanics Useful?" (*Philosophical Transactions of the Royal Society of London A* 353 [1995], pp. 367–76).

169 "Rather than teaching us how to build a large quantum computer": The article, rather dramatically titled "Quantum Computing: Dream or Nightmare?" by Serge Haroche and Jean-Michel Raimond, appeared in *Physics Today,* August 1996, pp. 51–52.

169 harder than NP-complete: See L. J. Stockmeyer and A. K. Chandra, "Intrinsically Difficult Problems" (*Scientific American* 240 [May 1979], pp. 140–59; reprinted in the book *Trends in Computing* [Scientific American, Inc., 1988]).

170 If there is none, it will churn and churn forever: Implications of the halting problem, as it is called, are explored in Douglas Hofstadter's classic book, *Gödel, Escher, Bach: An Eternal Golden Braid* (Basic Books, 1979).

171 the monks and the computer: Clarke's story can be found in a collection also called *The Nine Billion Names of God,* reissued in 1987 by New American Library.

A Note on the Notes

Only a handful of the papers in the vast and rapidly expanding literature on quantum computing have been mentioned here. For those who want to dig deeper, a good place to start is the University of Oxford's Centre for Quantum Computation, with the memorable address http://www.qubit.org. I could list many more, but the Oxford site is full of pointers that will soon have a visitor wandering all over the Web.

Those looking for more information can also turn to several good books, which assume various levels of mathematical acumen and patience/hunger for detail. In ascending order of difficulty, I would rank them as follows: Gerard J. Milburn's *The Feynman Processor: Quantum Entanglement and the Computing Revolution* (Perseus, 1998), Julian Brown's *Minds, Machines, and the Multiverse: The Quest for the Quantum Computer* (Simon and Schuster, 2000), and Colin P. Williams and Scott H. Clearwater's *Ultimate Zero and One: Computing at the Quantum Frontier* (Copernicus, 2000). I found these very useful as references and particularly found myself going back to Julian Brown's comprehensive overview. The bible of the field is Michael A. Nielsen and Isaac L. Chuang's monumental *Quantum Computation and Quantum Information* (Cambridge University Press, 2000). I found the introduction, particularly the commentary on the potential power of a quantum computer, very illuminating and refreshingly clear.

Acknowledgments

To help ensure that this book is as accurate as I can make it, I sought the counsel of several scientists and mathematicians who read parts, and in some cases, all of the manuscript: Joseph Traub, Seth Lloyd, Lov Grover, Bart Selman, Umesh Vazirani, David Wineland, Artur Ekert, Matthias Steffen, Peter Shor, and Gilles Brassard. In addition I interviewed or corresponded with a number of other researchers over the years, including Charles Bennett, Isaac Chuang, James Crutchfield, David Deutsch, David DiVincenzo, Manny Knill, Raymond Laflamme, Rolf Landauer, Christopher Monroe, Phil Platzman, John Preskill, and Wojciech Zurek.

To keep the account as clear as possible, I relied on the good advice of family, friends, and colleagues: my wife, Nancy Maret, her brother Douglas Maret, Tim Palucka, Julie Pullen, Christine Kenneally, Olga Matlin, Martin Bronstein, Deborah Blum, Rabiya Tuma, Dana Hall, and Steven Talley. In addition some of the participants in the 2002 Santa Fe Science-Writing Workshop offered insightful comments: Sara Robinson, Bridget Rigby, and Josh Winn. Thanks to them all, and apologies to anyone I have forgotten and for any errors that remain.

As always, I was cheered along the way by the enthusiasm of Jon Segal at Knopf and Will Sulkin at Jonathan Cape. And my agent, Esther Newberg, did her usual fine job of negotiating the contract. I especially want to thank the book's designer, Peter Andersen; its production editor, Kathleen Fridella; and its illustrator, Barbara Aulicino, for doing such a beautiful job turning my 2.7 megabytes of 1s and 0s into something solid.

This is the first book I have written that my father will not be able to read. It is dedicated to his memory.

Index

Page numbers in *italics* refer to illustrations.

A Note on the Type

This book was set in Minion, a typeface produced by the Adobe Corporation specifically for the Macintosh personal computer, and released in 1990. Designed by Robert Slimback, Minion combines the classic characteristics of old style faces with the full compliment of weights required for modern typsetting.

Composed by North Market Street Graphics, Lancaster, Pennsylvania. Printed and bound by R. R. Donnelley & Sons, Harrisonburg, Virginia
Illustrations by Barbara Aulicino
Designed by Peter A. Andersen